in situ
尹西图事务所 设计作品专辑

public landscapes
开放景观

Green Vision 绿色观点·景观设计师作品系列

本系列图书为法国亦西文化公司(ICI Consultants/ICI Interface)的原创作品，原版为法英文双语版。
This series of works is created by ICI Consultants/ICI Interface, in an original French-English bilingual version.

法国亦西文化 ICI Consultants 策划编辑

总企划 Direction：简嘉玲 Chia-Ling CHIEN
协调编辑 Editorial Coordination：尼古拉斯·布里左 Nicolas BRIZAULT
英文翻译 English Translation：柯尔斯顿·薛帕尔德 Kirsten SHEPARD
中文翻译 Chinese Translation：王美文 Mei-Wen WANG, 简嘉玲 Chia-Ling CHIEN
版式设计 Graphic Design：维建·诺黑 Wijane NOREE
排版 Layout：卡琳·德拉梅宗 Karine de La MAISON

绿色观点·景观设计师作品系列

green vision

in situ
尹西图事务所 设计作品专辑

emmanuel jalbert
埃曼纽尔·加勒贝尔

public landscapes
开放景观

广西师范大学出版社
·桂林·

images
Publishing

献给 艾芙·玛尔

法国国家景观设计师与建筑师，曾任本事务所项目设计经理。由于医务上的失误，她在捐赠血浆的过程中丧失了生命，卒于2009年9月29日，仅享年29岁。

如今，她曾经设计过的场所、完成的方案、种下的树木，都各自展现着生机。这些景观以沉默的方式向她过于短暂的生命致意，虽如昙花一现，却传达出喜悦、慷慨、充实……

埃曼纽尔·加勒贝尔

To Eve Marre

*DPLG Landscaper, DPLG Architect, Project Leader,
who died at 29 years of age, on 29 September 2009,
after a medical error during a plasma donation.*

*The places that she designed, the projects that she seeded,
and the trees that she planted, live their lives today.
All these landscapes render a silent homage
to her too short existence, joyous, generous, intense...*

Emmanuel Jalbert

右图：位于圣克里斯托尔、里昂、舍诺夫等城市的整治空间
Opposite: In Saint Christol, Lyon, Chenôve

contents 目 录

创造宜人的生态环境
environments of life — 007

空场与边界
clearings / border areas — 010

实用与邻里
uses / neighbourliness — 032

场所与环境
places / environments — 060

岸地与河流
banks / rivers — 084

方案索引
projects index — 124

事务所经历
experience — 131

团队
team — 132

活动与得奖记录
interventions-distinctions — 133

版权说明 / 致谢
credits / acknowledgements — 134

创造宜人的生态环境
environments of life

In situ（音译尹西图）：拉丁文片语，意指在当地、就地、处于自然环境中。在当代艺术中，in situ意味着在进行创作时，因地制宜的考量。

尹西图景观事务所由景观设计师与城市规划师等"空间匠人"所组成，致力于创造能够赋予城市活力的开放空间。二十五年来，我们实现了非常多样化的项目，包括从大尺度的地域规划到仅能容纳一棵树、一张长椅和过路行人的迷你花园。在我们眼中，项目无大小之分，重要的是对每个场所给与持久的关照，以简约却永续的方案来创造宜人的生态环境。

无论在此在彼，尹西图事务所都就地取材、因地制宜地进行设计，尽管脚踏实地，却也发挥着无限想象创意。对每一个基地的造访，都使我们得以发现其独特的性格、认识其所处的地域，以及与其不可分割的广大地理环境。我们亲临基地，用脚步跋量、走遍每个角落、测试坡度，同时也评估方位朝向、日照状况、土质优劣，并且感觉风向……在这些阔步探寻的过程中，方案构思逐渐诞生。我们眼观地平线尽头，以便将目光放远、领会最大尺度的空间；同时也超越基地范围，探索地势的界限与高低起伏，观察植物生长进程、水流曲线、行人路线及所有当地习俗。我们必须解读这些因素的互动关系与潜能，才能进行规划与设计。

In situ: a Latin expression that means in place, on the site itself, in its natural environment. In contemporary art, it signifies a work that takes into account the place where it is installed.

As landscapers and urban planners, artisans of spaces, the In Situ studio intervenes in the empty spaces that are the basis of the life of cities. For 25 years, we have had the pleasure of realising highly varied projects that range from the vast scale of the greater area of a city to that of a small garden that hosts just a tree, a bench and some passers-by. To our eyes, there are no large or small projects, but a constant attention to each place, in order to conceive simple and sustainable projects and to install "environments of life".

To be here, there, in situ, feet on the earth and head in the air. Visiting a site is always the discovery of an unique situation, of an area that is part of a much more vast geography. Pacing the expanse, that's where a project is invented: physically measuring a plot of land on foot, exploring it in all directions, checking its steepness, appreciating its sun exposure, the quality of its light, that of its soil, sensing the direction of the winds... The eye embraces the horizon in order to see the farthest and to apprehend the greatest dimension. Leaving the framework, exploring the limits, the inflections of the relief, reading the evolution of the vegetation, the water ways, the routes of the passers-by and all the customs of the place: there are so many interrelationships and potentialities to interpret in order to be able to plan.

左页：位于里昂、修蒙、罗歇塔耶、舍诺夫、马孔等城市的整治空间，展现出多样化用途

Opposite page: Different uses in Lyon, Chaumont, Rochetaillée, Chenôve and Mâcon

随后,则由手来接管下一步,手绘过程伴随着思路,更经常领先思路,一笔一画掌握各种线索,画下分水线、等高线、临界点、系统网络。这些绘图犹如描述地域风貌的手稿,记录了自然风土,也记载着历史、地理及文化的沉淀……逐渐地,方案设计从错综交织的草图线条中逐渐成形,先在纸上定位,进而落实到土地上。每个地域都受到强大的活力变数所影响,决不会是沉滞而无生机的。所有方案的组成元素通常都已存在于基地本身,每个地方都蕴藏着促使自身转变的因子。我们需要做的,便是强化、扩张和巩固已经存在的一切,并重新建立连续性,为特定情况顺水推舟。每个方案都从所有"既存"的一切出发,而设计的关键则在于知道如何展现这些基地元素,并加以串连、组合、重缀、修整、恢复活力……进而得以实现项目的宗旨。整治空间也意味着对既有状况的处置,虚心融入万物合奏的乐章中,不仅做好整理的工作,同时也辟出空间、直入核心课题并赋予意义。我们的一切方案设计皆试图朝向更合宜、简约的方向迈进,不断地寻求"以最少的资源达到最佳的效果"。

生物至上。一切景观的首要"原料"便是生物:植物、水、腐殖土、动物和人类,这些都是生态系统的一部分。城市是具有生命力的有机体,而我们日常所见的景观则构成了人类的生活环境,此环境不断地迈向城市化状态。这个观点强化了我们为城市空间注入活力的决心,使其能够同时容纳自然和人类。在每个项目设计中,我们都优先考虑生物与居民的需求,将雨水花园、生态沟渠、草场、林地、树丛等元素与建筑周围的花园、儿童游戏场、公园及行人步道等空间紧密结合,同时也将其融入市集林荫道和城市广场的设计中。我们为"大型居住区"(法国1950-1970年代的社会住宅)重新规划家庭式花园或分享式花园的各种经验,都属于这种兼顾人文与自然的理念和方法,由此创建一个有利生态发展的环境,并促进社会凝聚力。

Then the hand takes over, it accompanies and often proceeds thought, explores, traces the watershed, the contour map, the thresholds, the networks. Manuscripts of an area, these maps record nature's share and the share belonging to the historical, geographical and cultural sedimentations... Little by little, the project is drawn, in the interlacing of the strokes of the sketches, they slowly take root on paper, then in the density of the soil. An area is never inert, it is animated by dynamic powers at work. The defining elements of every project are very often contained in the site itself, each place carrying in itself the seeds of its transformation: the work then becomes amplifying, propagating, supporting that which exists, re-establishing continuities, becoming an actor in a situation. Each project builds on everything that is "already there". The challenge consists in knowing how to reveal it, then to re-connect it, re-unite, re-sew, repair, reactivate and realize. Developing thus means also knowing how to arrange what exists, modestly entering into the harmony of things, cleaning house, freeing the space, going to the core and giving it meaning. Each project must tend towards more relevance and simplicity, always trying to do "to one's best, with less".

Concerning the living. The essential primary resource of every landscape is the living: the trees and the plants, the water, the humus, the fauna and also all the people who are an integral part of these eco-systems. Cities are living organisms and our daily landscapes form existential environments of human beings, more and more urban. This point of view feeds the desire to enable places to host both nature and humans. At every occasion, priority is given to the living and its inhabitants: rain gardens, ditches, meadows, woods, groves of trees are closely associated with the gardens around buildings, with playgrounds, parks and routes, but also with the mesh of the market as well as town squares. Our various experiences of family or community gardens at the heart of large ensembles participate in this hybrid process that creates a fertile environment that both feeds people and reinforces social ties.

时光造化。自然景观诉说着光阴岁月的造化，也展现时间流逝的痕迹，犹如以大地为羊皮纸稿，一再让前文隐迹之后又重新写就的篇章；自然景观具有持续演变和转化的杰出特点，因此能穿越时光而恒存。每个场所都带有记忆，对于能够辨识此记忆的人而言，为一个场所进行整治规划，意味着对其历史进行了解，以便预测它日后的变化，将其过去、当下和未来串联起来。基于对持久性的关注，我们采取结合了近程、中程和远程目标的景观策略，规划出具有演化能力的发展进程，在其中纳入了施工与短暂活动的时间，以及白天和夜晚、四季更迭、植物生长的时间……而植物通常是不会停滞不前的！以规划来伴随景观的转变，意味着对方案实现后的阶段也赋予关注：预测方案的持久度与远期效果、设想其未来发展与管理模式。大自然与景观的时间缓慢悠长，而仅是过客的人类的时间却短暂而紧促，我们经常得为这两种不同的时间尺度找到和谐的步调。

开放景观。公共椅凳、公共过道、公共空间……城市景观在更大的尺度上具有名副其实的公共使命：河畔、自然空间、街道、广场、公园，这些属于每个城市居民的空间形成了绿色网络、开放的路径、流通的空气，以及城市肌理中的连续性元素。它们并非城市的"布景"，而是不折不扣的生活场所，是人们共同生活的联系与共享的资源。城市发展区域总是越来越拥挤、受到各种规划的限定、呈现片段状态，同时也更加封闭孤立，并走向私有化，而自然景观则提供了人们绝佳的喘息休憩场所，这个自由的空间向多元化的大众敞开，为所有公民所共享共有。此概念同时表达了我们对景观规划的观点以及设计伦理，即致力于推广具有高度立场的城市政策，以创造高质量、具有连续性且自由开放的公共景观。

With time. Landscapes speak to us of time and of seasons. Terrestrial palimpsests, they have the great virtue of evolving and transforming without cease in order to endure. All places have a memory, for those who know how to see it and the act of intervening on a place implies understanding its history in order to envisage its future, in a continuum that links past, present and future. This constant taking into account of the span of time leads to the implementation of landscaping strategies that join the short, the medium and the long term. We project evolutive processes that integrate the time of work and construction, that of ephemeral events, that of days and nights, of seasons, the time of vegetation… which more often than not do not vegetate at all! Accompanying the transformation implies anticipating, observing the resiliencies, managing expectations and imagining the future and the maintenance of its realisations, beyond the mere delivery. It means very often reconciling the slow and long time of nature and of landscapes and the much shorter and more urgent time of humans, who are just passing through.

Landscapes open to the public. Public benches, public passages, public spaces… On the largest scale are the landscapes of the cities that have a veritable public vocation: river banks, natural spaces, the networks of roads, town squares, and parks belong to everyone, they make up a green mesh, open routes, currents of air, elements of continuity in the urban fabric. They are not "décor", but places of life in themselves, connections and common assets to use and to inhabit together. At the heart of urban agglomerations that are always more encumbered, programmed, fragmented, closed in and privatised, all these public landscapes form a breath of fresh air, spaces of liberty, open to diversity, commonality and citizenship. This notion expresses both a viewpoint and an ethic of a process, of a city's policy engaged in developing the quality, the continuity and the liberty of landscapes open to the public.

10

"边界是一种特殊的空间形态，
各种关系间的密切张力在这里显得更集中、更显明。"
米歇尔·寇拉儒

"A boundary is a specific form of space where, more than anywhere else, the intensity of relationships is concentrated and made manifest"
Michel Corajoud

clearings / border areas
空 场 与 边 界

在森林深处，一如在密实的城市肌理中，空场构成了相当独特的景观形态：这些充满着空气的虚空间与土地和光线紧密相连；其边界形成一个天空框景，并且构成一个丰富且密集的生态环境。城市广场便犹如林中空场，形成一种虚与实之间的张力，一种内容物与容器之间的对话，并产生如剧场空间般的特点：中央开阔的空间经常为"演员"所使用，而"观众"则更常靠在边界地带——露天咖啡座上或建筑立面墙下。这种开放空间的设置不仅具有象征性意义，也成为居民交往的社会性场所，尹西图事务在许多不同项目设计中都加以深入发展。同样地，对抗社会空间的拥挤、封闭和隔离现象也是尹西图事务所始终如一的设计态度，其经手整治的空间必须兼具开放性格和社会责任。我们不该对"空"有所恐惧，而是应该懂得如何为它留出一席之地，并加以定位、塑造，使城市生活和人际往来都能在其中自由开展。

In the depth of forests, as in the density of our cities, clearings are unique features of the landscape: bare spaces, these airy volumes unite earth and light. The border area around clearings frame the sky and form rich and intense environments. Town squares, essentially urban clearings, create an analogous tension between filled and empty space, a dialogue between the contained and the container that evokes a theatrical set: the open central space often hosts the "actors", while the "spectators" occupy the border areas, at the cafe terraces or at the feet of building facades. This open spatial arrangement, conveying both meaning and social practices, has been developed through several projects of the In Situ studio. Similarly, combating the problems of encumbrances, confinement and segregation is a constant of the studio's projects: their developed spaces are both open and maintained. One should not have fear of the "empty", but rather know how to make it a place and shape it, so that life and all its social activity can freely unfold.

法国 维勒班 / 1997

the gardens of Pélisson city
佩里松小区花园

花园城市的居家景观

位于里昂郊区的佩里松社会住宅小区内,四处都是随意停放的车辆,而其外部空间也破败失修。限于拮据的工程经费,尹西图事务所对这些空间的整治只能进行重点处理,无法重新考量现存道路系统与设备网络。方案以重新改造每个空间为要务,透过功能和景观的结合来重新建构一个"花园小区"。

街道经过重新调整,以方便停车及栽种樱花树。中央草地被重新整治为下沉式空间,而从中挖出的土方则被堆在周围,形成围堤,并在其上栽种不同的植物。这个混合处理的边界地带为此空场建立了框架,中央成为嬉戏与街区节庆活动的场所。小区之家的中庭、儿童游戏场、小区前广场等其他开放空间则成为用途各异且具有互补氛围的城市空间。每个场所都界限分明,容易辨识。在两栋长条形、不设有阳台的建筑之间,居民们利用中庭来晾晒衣物。尹西图事务所建议强化这一用途,设置了一个长形晒衣棚架。这个住宅花园兼顾了实用与舒适的面向:小区居民会聚于此,在芳香扑鼻的白色紫藤花下一边晾衣服,一边闲话家常。

这个早在1995年便已实施的项目显示出事务所在公共空间的设计上重复出现的主题:以最少的经费赋予每个空间价值,透过植物栽种、实际功能(尽管是十分平凡的用途)的改善,以及对场所形态的关注,来促进邻里和睦。

此方案荣获1997年法国景观锦标奖。

The domestic landscape of a gardened estate

On the periphery of Lyon, the low-income housing estate of Pélisson was invaded by cars, and its exterior spaces were much deteriorated. The very small construction budget led In Situ to keep itself to the essentials, without changing the lay-out of the existing streets and the transportation network. The priority was thus given to the renovation of each space through work divided between public use and landscape, in order to create, in retrospect, a "gardened estate".

The streets were re-graded to make space for parking and for the planting of ornamental cherry trees. The central meadow was slightly dug out and the backfill moved to the perimeter to form a small embankment, lined by a rim of various plantings. This mixed border drew a frame around the little clearing used for games and neighbourhood parties. The courtyard of the social centre, the playground and the parvis formed other urban "rooms" with complementary uses and ambiences. Each place was delineated, and identifiable. Between two long buildings, equipped with balconies, the residents dried their wash in the courtyard. In Situ proposed to reinforce this custom by installing a long clothes-line trellis. This domestic garden marries the useful and the pleasant: the inhabitants of the neighbourhood come together to hang laundry and to chat under the white and perfumed wisteria.

This project, started back in 1995, is a manifesto of In Situ's recurrent working themes in regards to public spaces: the desire to renovate each space at the lowest cost, the choice to restore a sense of dimension through planting, the preservation of customs as modest as they might be, and the attention given to the configuration of the spaces, in order to promote neighbourly relations.

The project received the national prize of the Landscape Trophy in 1997.

上两排图：晒衣棚架 / 施工前基地实景，设置于建筑物边上的晒衣架
下两排图：桦树林 / 中央草坪 / 边界地带的绿化

Top: The trellis – clothes line / View of the site before work, the clothes lines in front of the buildings
Bottom: Birch plantings / The central meadow / The planted border

上图：晒衣棚架
下图：中央草地

Top: The trellis
Bottom: The central meadow

法国 里昂 / 1998-2002

8 may 1945 place
1945年5月8日广场

混合式公共空间：
市集广场与公共花园

1945年5月8日（第二次世界大战欧洲胜利日）广场是里昂的第二大广场，邻接东尼·加尼叶设计的美国街区。这个空旷的广场位于人道边，草木不生，占地达3公顷，曾提供大型马戏团的使用。早期每周一次的室内市场设在广场平台一角的仓库里，然而，这个备受欢迎的菜市场却是整个街区真正的经济动力所在。尹西图事务所建议突显这项重要用途，在沿着大道的广场边缘设置一个全新的市集厅棚，成为既有城市肌理与商店的延伸。

建筑师弗朗索瓦兹-艾莱娜·朱达在此建造了一个架在三排去皮原木桩上的长顶棚作为市场用地，形成广场和大道之间活跃的过渡空间。广场北方边界处是长条状的住宅建筑，建筑脚下宽阔的带状空间种植了绿色植物，并在松树和紫荆树荫下设置了儿童游戏区和滚球场。停车场被设置在西边的绿化平台内，这里同时也作为每周六织品市集的延伸使用空间。稍微下沉的中央空间形成一个大型绿地"舞台"，面积达5,000平方米。

这个"草地广场"是进行球类运动、表演等活动的场所，周围的阶梯和缓坡强化出此开放空间的界限。在花园区，挡土矮墙上点缀着喷水，水柱喷出后流入线型水池中。这个混合式的公共空间极为特殊：它同时是市集广场、公园和游戏场。边界地带的植物和建构物形成了中央空场的框架，构成一个提供市民交流共享的空间。

此广场设计荣获2003年Le Moniteur出版集团城市整治奖。

A hybrid public space:
market place and public garden

Lyon's second biggest public square, the 8 May 1945 Place adjoins the United States Neighbourhood conceived by Tony Garnier. At the edge of the boulevard, this sterile space of three hectares hosted large circuses. A weekly covered market housed in a simple warehouse was positioned in a corner of the square. However this very popular farmer's market was always the economic engine of the neighbourhood. In Situ proposed to highlight this important activity by installing a new market hall at the edge of the public square, along the boulevard, in an extension of the urban fabric and the existing businesses.

Playing with the vocabulary of the traditional tree-lined promenade, Françoise-Hélène Jourda built a long canopy supported by three rows of posts of raw, stripped wood, which became an active filter between the area and the boulevard. At the foot of the apartment buildings, the north edge of the public square is devoted to a wide planted strip, equipped with playgrounds and bowling greens, under the foliage of pines and Judas trees. At the west, a planted esplanade shelters parking places and the extension of the Saturday textile market. The central space, positioned at a slightly lower level, forms a grassy green arena, a sort of "sunken" reinforced lawn of 5 000 m²

The "public lawn" hosts ball games, open-air shows, etc. The stairs and ramps delineate the limits of this sunken grassy rectangle. On the garden side, the low retaining wall is punctuated with fountains that pour water into a linear basin. This atypical public space is hybrid: at once market place, public garden and ball field. The planted and constructed borders frame the central clearing, to create a space of diversity and of commonality.

This town square received the Trophy of Urban Development of the Moniteur 2003.

左页上图：竞赛方案 / 施工前鸟瞰照片
右页上图：大草地和市集厅棚
下排连图：广场及其空间用途——儿童游戏场、大草地、休憩场所、喷水游戏区……

Left: Competition submission / Aerial view before work
Right: The meadow and the hall
Bottom: Detailed view of the square and its customary uses: sports fields, large meadow, areas for relaxation, water fountains...

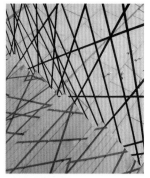

市集厅棚和其边缘地带，一个日常生活场所：市集活动前、中、后的空间使用和细部处理（弗朗索瓦兹-艾莱娜·朱达设计的市集厅棚）

Mosaic "the market hall and its surroundings, a lively place", detailed views of the developments and uses of the site, before, during and after the market (The market hall, created by Françoise-Hélène Jourda, architect)

上图：开放的大草地，市集厅棚成为草地与街区之间的过渡场所；装设照明设施的大型桅杆
下图：广场夜间照明设计，蓝色光线流入草地上，形成游泳池般的视觉效果

Top: The large open meadow, the market hall creates a transition with the neighbourhood; view of the large masts used a lighting fixtures
Bottom: Lighting of the square, a blue swimming pool fills the meadow below

上图：从绿化边界地带看向中央大草地（松树和紫荆树，搭配岩蔷薇和薰衣草草棉）
下图：下沉式中央草地，位于边缘的喷泉与水柱，远处为绿化停车场

Top: View of the meadow from the planted borders (pines, Judas trees, carpet of rockroses, lavender cotton)
Bottom: View of the "sunken" meadow, the fountain and the water jets

左图：在槐树与漆树凉荫下谈天说地
中图：大草地上的球赛，一个日常的游戏空间
右图：带状水池

Left: Discussions in the shade of sophoras and sumac
Middle: Ball game in the big meadow, used daily as a sports field
Right: View of the linear basin

瑞士 洛桑、勒南、普里伊 / 2011-2015
the corridors of Malley
马利街区平行绿化带

以煤气库公园为中心的可持续发展街区

位于洛桑西部的马利街区是城乡发展的规划基地之一。众多旧屠宰场和工业用地迁移后，这里成为新兴活动的天地，并得以建设临近火车站的住宅区，这一生机勃勃而肥沃的街区占地将近50公顷。

尹西图景观事务所与FHY建筑与城市规划事务所联合赢得2011年公开设计竞赛的首奖，提出一个循序渐进的转化计划，强化虚实空间的关系，也将建筑物和景观结合在一起：以一个植物围篱系统来建立起基地的景观结构。这些犹如五线谱的平行围篱以时断时续的线条滑进了建筑物之间，在公共空间和私人空间之间创造出具有连贯性的植树环境。街道、公共设施、建筑群和公共空间逐渐定位。穿越街区的通道形成了一个由慢行交通构成的次要网络。道路网络连结地面排水渠道，将基地的雨水导向一个池塘。煤气库公园围绕着这个池塘而展开，水面映照出基地的象征标志"球形煤气库"。公园的大草地成为整合整个街区的中央空场，而公园的边缘地带则设置了共享果园与花园。

这个进行中的项目分为短期和长期规划。我们提出的方案策略能快速完成某些建设，以便迎接首批居民。火车站广场、公园和某些其他广场的景观都呈现了雏形，并将在日后持续发展完备。"行远必自迩"，这个简约、灵活而具有演化能力的规划为城市提供了长期发展的潜力，不仅可以避免因一时的过度规划而使计划触礁，同时也能够让每个公共空间的功能与属性拥有随着时间推移而调整的可能性。

A sustainable neighbourhood around the Park of the Gazomètre

To the west, the Malley Neighbourhood is one of the urban development sites of the conglomeration of Lausanne. The removal of the slaughterhouses and various industries has made way for new activities and residences near the train station, in this newly fertile neighbourhood of more than 50 hectares.

The winners in 2011 of an open competition, In Situ and FHY proposed a progressive transformation that combines work on empty and filled spaces, the constructed and the landscape. A system of vegetal corridors forms the landscaping framework of the site. Like a musical stave, fragmented parallel lines slide between the buildings and create continuities of plantings between public and private spaces. Roads, facilities, constructed city blocks and public spaces are placed progressively. Paths across the area develop a secondary network of soft means of transport. The road network is lined with surface gutters that lead rainwater to the pond. The Park of the Gazomètre unfolds around a reflecting pond and the emblematic "gas bubble" reflected in it. The big meadow of the park becomes the central clearing that unites the entire neighbourhood. Community orchards and gardens have been developed as borders for the park.

This project is part of a short- and long-term study. The proposed strategy makes possible rapid implementation of some of the structures, to receive the first residents. The landscaping of the parvis of the train station, the park and some of the squares are designed for subsequent completion. Step by step, this serious, flexible, and evolving urban process marches towards durability. It avoids the pitfalls of over-programming and leaves room for manoeuvring in order to adjust the purpose of each public space over time.

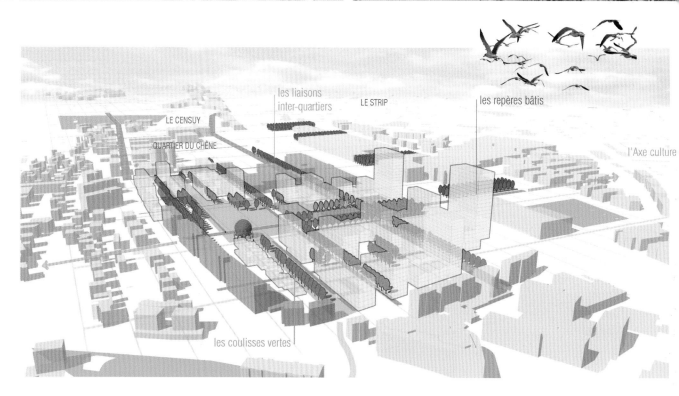

上排图：上个世纪初建立的马利煤气工厂 / 煤气库公园 / 水池和球形煤气库
下排图：平行配置的植物围篱 / 马利街区总平面图

Top: The Malley Gas factory at the beginning of the last century / Park of the Gazomètre / The pool and the Gas Bubble
Bottom: The vegetation corridors / General plan of the Malley Neighbourhood

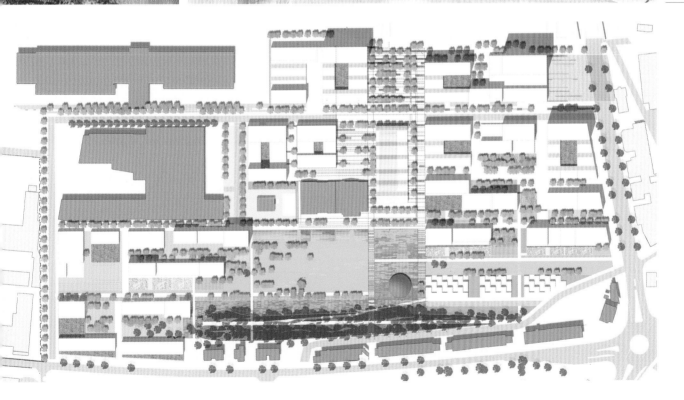

法国 伊塞尔河畔罗芒 / 2014

the great promenade of Vercors
韦科尔大型散步道

从林荫大道到散步道，从广场到城市花园

罗曼镇的让-饶勒斯广场和周围的林荫大道从西向东伸展，形成旧城区和外围街区之间的广大隔离带。这个隔离带如今却具有将空间重新"缝缀衔接"的使命。罗曼镇的城市结构清晰易读，旧城区具备许多可利用空间、排列整齐的行道树、丰富的建筑文化遗产，以及根据日照方位与阴影而配置的实用空间与人行路线。本方案设计便是以这些优势为基础而展开：采用"既存的"且运作正常的一切元素，来强化每个场所的优点与特质。

这条大型散步道的整治使得一条广阔而商店林立的绿化人行道可以获得延续，并且减缓交通流量，为行人与自行车族带来更大更舒适的空间。方案通过重新安排南北向路径，来重建上下城市之间的关系。让-饶勒斯广场的中央三角形空间拥有远眺韦科尔山脉的视野，并且可容纳露天市场、集会游行和各种特殊活动。这个可因应需求而改变用途的公共空间，还时而变化为地面水镜与喷泉。这个整治方案旨在建立城镇中心和外围街区的关系，并发展丰富多样的空间使用方式，以及建立充满活力的整合性公共空间，提供市民和谐共享的场所。

这个经过长时间咨询的竞赛得奖方案在最新一轮市长选举后无疾而终。新任市长更偏好在旧城中心设置许多停车场。然而，罗芒这个著名的制鞋之都，难道不更应该提供人们"徒步"游走而非车辆使用的空间吗？

此方案荣获2014年咨询竞赛首奖。

From the boulevards to the promenade, from the town square to the city garden

In Romans, the boulevards and the Jean-Jaurès Place stretch from west to east, and form a gaping boundary between the old centre and the neighbourhoods. However, this tear in the urban fabric is also an opportunity for a true alteration. Romans possesses a readable urban structure, lots of available space in the middle of the city, aligned trees, an architectural heritage, as well as uses and itineraries resulting from the orientation of the sun and the shade. Many of the area's positive qualities can become the foundation of the project: that which is "already there" and which already works, through strengthening these qualities and the identity of each of these places.

The development of a large promenade assures the continuity of a big, planted and shady footpath alongside businesses. This rearrangement settles the traffic down and gives space to and promotes the safety of pedestrians and cyclists. The links between the upper and lower city are re-established by re-organizing the north-south crossings. The central triangle of the Jean-Jaurès Place has a great view of the Vercors and can be used for open-air markets and public demonstrations… A reflecting pond and water works ornament this multi-faceted public space. The major thrust of this development is to reconnect the centre and the neighbourhoods, to develop diversified uses, to create living and unifying public spaces, places for activity and community.

This project, winner of a long competitive dialogue was not pursued in the end, after the last municipal elections. The new mayor prefers to develop several parking structures in the interior of the old centre. Should not Romans, famous for being the capital of shoes, be travelled by foot rather than in a car?

Winner of a competitive consultation in 2014.

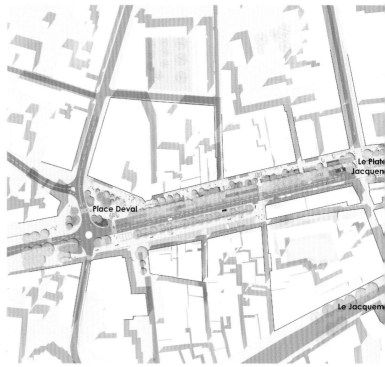

大型散步道的雨水管理原则：四分之一的渗水地面（绿化+透水性铺地）可以避免雨水流到排水道

Principal of the management of rain water on the great promenade: 1/4 porous surface (plantings + filtering coat) enables the limiting of waste in the network

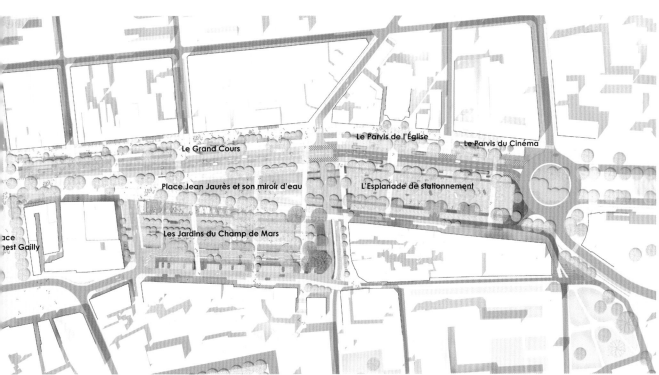

上排图：大型散步道透视图 / "练兵场"花园透视图
下排图：替代式雨水管理方法的原则示意图 / 大型散步道平面图

Top: Perspective view of the great promenade / Perspective view of the gardens of the Champ de Mars
Bottom: Diagram of the principle of alternative rain water management / The great promenade

"我们景观设计师是空间的行动者。
我们烙下痕迹，赋予事物意义，这才是真正必须分享的……
我不是在打造城市艺术或城市小品，我创造的是人性化的场所。"
吉勒·维克斯拉德

*"We landscapers are the actors of space.
We imprint, we give a sense to things that's what we must share.
I don't make urban art or urban objects, I make human places."*
Gilles Vexlard

uses / neighbourliness
实 用 与 邻 里

通俗的"绿化空间"概念倾向于将这些场所简化为色彩统一而贫乏无味的"叶绿素布景"，在此之外，每个景观确实是由复杂、丰富而多样的生态环境所组成，紧密连接了土地、植物、野生动物和包括人类在内的各种其他生物，而人类每天都在创造与使用景观环境。我们的城市景观首先必然是实用空间，而"真正的"景观，无论从植物学角度，或从生态、地理和社会意义而言，都是一种建立联结与关系的艺术："公共空间"是人们互相交流、和谐共处的必要场所。通过景观的媒介，空间才能实际具有共享的特质，也才能将地域的主要特质展现出来。成功整治的空间，首先便必须是生气蓬勃的场所，是一个允许多种不同风格、条件以及不同世代交融共处的多元化空间，并且有利于建立良好的邻里关系。尹西图事务所试图通过每个经手的项目，来促进各种文明和谐的人际往来，创造良好的空间联系与社会交流，以赋予每个场所实质的意义。

Beyond the common notion of "green" spaces, which tends to reduce these places to a uniform colour and a mere "chlorophyllian décor", each landscape is fully composed of living environments, each complex, rich and varied, intimately linking the earth, vegetation, the fauna and many other beings, among them humans, who use it daily. The landscapes of our cities are above all spaces that are used, and "The" landscape is an art of connection and relationship, in the botanical sense as well as the ecological, geographical, and of course the social sense: "public spaces" are essential places of common ownership and citizenship. Space there is de facto shared and therein lies the principal characteristic of these areas. A successful development is above all a place that lives, a space of diversity that permits the blending of genres, conditions, as well as generations, and that facilitates good neighbourliness. Through each project, In Situ tries to promote all connection and all civilities, to give meaning by creating both spatial and social links.

法国 沃昂夫兰 / 1996-1997

La Thibaude neighbourhood
提 伯 德 街 区

塔楼下的共享花园

沃昂夫兰镇位于里昂市郊，是1960年代法国城市规划中设定的大型优先发展区，并以此而闻名。提伯德街区位于这批大型社会住宅群中，年久失修而严重损毁，而且四周空间封闭，整个社区犹如密不透风的堡垒。大里昂都会与法国低租金住房管理机构合作推出了市区更新政策，降低了本街区的密度，并重新规划地面使用场所，带来焕然一新的街区面貌。这个建于30年前的街区，从此将呈现崭新而生机盎然的城市景观。

1960年代注重经济效益的建设逻辑影响了这些住宅塔楼的配置，也因此造成大面积的阴影区域，其中三栋塔楼已经拆除，整体建筑群经过重新修复后都漆成白色，有些地下车库也已拆毁作废。这里终于能再度看到阳光和天空，也能从释放出的空间设计一条横向街道、一个中央广场和一些植树停车场。地产权经过重新分配以后，规划出三个可辨识的居住街坊，皆具备精致的围墙和篱笆。住宅区花园种植了桦树，并设置了儿童游戏场，而以几个岛状空间来种植欧洲赤松的中央广场则提供了许多其他使用功能。

这个社会住宅群的南部边界与别墅型住宅区相接，过去由一长排空心砖堆砌的墙面来区隔，上面还架设带刺的铁丝网，呈现出相当严肃的面貌。尹西图事务所建议在此设置带状蔬果园，让这个边界显得更友善、更具生命力，而过去这两个互不往来的区域却从此建立了和谐的邻里关系。在这个成效卓著的经验之后，河岸附近的街区也因此诞生了许多其他的共享花园。

Community gardens at the foot of residential towers

The municipality of Vaulx-en-Velin, in the suburbs of Lyon, is known for its vast ZUP (priority urban development zone) from the Sixties. Among these large complexes, the very deteriorated La Thibaude Neighbourhood strongly resembled a fortress, and suffered from its inaccessible location. The policy of urban renewal launched by Greater Lyon and the HLM offices worked to render this neighbourhood less dense, to give it gardening possibilities and to improve its image. Thirty years after its construction, this neighbourhood now benefits from a new, alive, urban landscape.

Arranged for cost-effective construction around the paths of cranes, the residential towers generated large shaded areas. Three of the towers were demolished and the entire complex was rehabilitated and repainted white. Some of the partially underground parking structures were also demolished. All these changes permitted the neighbourhood to rediscover the sunlight and the presence of the sky, but also gave them the free space to create a transversal road, a central square and planted parking areas. The redrawing of property lines delineated three readable residential blocks, equipped with well-made fences and hedges. The residential gardens, planted with birch, are equipped with playgrounds, while the central square, punctuated with little groves of forest pines, is used for many different activities.

On the southern border area of this big complex, the very hard boundary with the nearby residential area is nothing but a long breeze-block wall, topped with barbed wire. In Situ proposed to render this border more pleasing and fertile by installing a strip of vegetable gardens. This plan helped to re-establish an easy neighbourly rapport between the two entities that had turned their backs on each other. Following this fruitful experience, other shared gardens were created in the midst of river neighbourhoods.

上排图：花园过道
中排图：中央广场
下排图：绿树荫下的过道

Top: View of the path of the gardens
Centre: the central square
Bottom: a path under the trees

基地全景,共享花园、广场和街坊花园

View of the entire project, the shared garden, the square and the gardens of the city block

法国 维勒班 / 2004-2008

Lazare-Goujon place
拉扎尔-古戎广场

摩天大楼街区中心的城市绿洲

拉扎尔-古戎广场位于维勒班市的摩天大楼历史街区，介于市政厅和国立人民剧院之间，广场的整治计划为这个具有浅色石灰岩铺地的广大徒步一古里新建立了基体性，成为一广阔的空场、一个城市绿洲。

原广场由1930年代的摩天楼建筑师莫里斯·勒胡所设计，本方案以当代手法延续原广场的组成元素，并加以重新诠释，尤其是两个方形水池和柱廊。南北两条街道的撤除使行人徒步平台得以拓展到建筑物的脚下，而广场下则建造了公共停车场。南方和北方两个边界分别植树的措施创造出一些平行的植物围篱，使广场尺度变得更具亲和力。市政厅前面的合欢树绿荫清幽，为岩蔷薇与玫瑰花园提供了遮护。在南边，国立人民剧院边上的皂荚树散步道则是人们闲逛和举办活动的绝佳场所。严谨的地面线条引导这个花园广场向两边道路敞开，而既有的两条柱廊经过延伸利用，形成了两座藤本玫瑰棚架，并为这个城市厅堂提供美妙的框景。

广场中央的两个方形水池仿佛镶嵌在地上的明镜，池水清浅，其水面高度与周围石基相齐，而池底则铺设了经过抛光的深蓝色砂浆水泥。由艺术家菲利普·法维耶设计的"天上园圃"将彩绘玻璃嵌入砂浆水泥，在池底形成迷人的星群图案。几道随机喷洒的水柱为水面激起涟漪，增添活力与生趣。无论是夜里或白天，这两个水池镜面都映照出摩天大楼顶上的天空……

An urban oasis in the heart of the Gratte-Ciels (Skyscraper) Neighbourhood

In the middle of the historic Gratte-Ciels Neighbourhood of Villeurbanne, between the Town Hall and the Théâtre National Populaire (People's National Theatre), the redevelopment of the Lazare-Goujon Place re-establishes the unity of a gracious white sandstone plateau for pedestrians. This vast clearing is an urban oasis.

This project extends and reinterprets in a contemporary manner the integral elements of the initial town square, designed in the 1930s by the skyscraper architect Môrice Leroux. The two square basins and the porticos received particular attention. The elimination of the two streets to the north and south permitted the extension of the pedestrian plateau up to the facades of the buildings, while a public parking structure was built underground. To the north and south, a more intimate scale is delineated by two planted borders which also create corridors of vegetation: at the foot of the Town Hall, the greenery of albizias brings a little shade to shelter gardens of cisti and roses. To the south, by the theatre, a mesh of honey locusts becomes the ideal place for strolls as well as temporary events. On the ground, the rigorous design ensures that this landscaped square opens to the lateral streets. The two existing porticos have been extended to form two trellises of climbing roses that frame this urban space.

At the centre of the Lazare-Goujon Place, two square, shallow, reflecting pools, are set into the ground, and the water is about on level with the stone base. The concrete basins are coated with a polished night-blue coloured epoxy "surfazo". The artist Philippe Favier has designed a "celestial garden" where constellations of painted glass imbedded in the surface, are strewn about the bottom. A few spouts of water fleeting and aleatory temporarily animate the skin of the water. Night and day, the two pools reflect the skyline… of skyscrapers.

左页上图：基地整治前后实景
右页上图：广场鸟瞰照片
下排连图：广场有多种用途，是一个供人休憩、交流、游戏、举办活动……的场所

Left: Before and after views of the site
Right: Aerial view of the square
Bottom: Different functions of the square, a place for relaxation, for encounters, for games, for the hosting of events...

« Tel un archéologue tête en l'air, je rassemblerais des centaines d'éclats bleus et or, comme autant de vestiges d'une nuit étoilée étrangement brisée.
Je tenterais de reconstituer ce zodiaque céleste, un "brin" enchanté ; rêvant que nos imaginaires s'inventent les mystères de ce puzzle à tout jamais éparpillé. »

"Like an archaeologist with my head in the clouds, I would bring together hundreds of blue and gold shards, like so many vestiges of a starry night strangely shattered. I would try to piece together this celestial zodiac, an enchanted 'wisp': dreaming that our imaginations invent the mysteries of this puzzle, forever scattered."

Philippe Favier, artiste/artist

上图：水舞，为水池带来生趣与活力
下图：市政厅绿廊繁花盛开，沿着绿廊设置的石灰岩长凳提供人们在合欢树的庇荫下休憩聊天

Top: The ballet of the water jets: playful activity of the basin
Bottom: Under the cover of albizzias, with bench-borders in limestone, the flowered corridors of the Town Hall

45

法国 马孔 / 2007-2014

the park walk of Marbé
马尔贝街区散步道公园

住宅街坊、林荫步道与共享花园

孤立而封闭的马尔贝街区在近十年来成为一项重大城市更新计划的整治项目。其既有的条状建筑与塔楼建筑像是被搁置在一个广阔的停车场上，然而，这个社会住宅社区的环境和可利用空间却拥有绝佳的优势。尹西图事务所竞标获选，负责进行公园和部分街道与街坊的规划设计。

这个城市规划以重新组织私人空间与公共空间、建立住宅街坊的穿越性街道网络为基础。拆除了部分塔楼之后，大幅拓宽了开放性视野，并得以展开一座具有横越性的带状散步道公园，在其中设置儿童游戏场、运动场和雨水花园。穿越公园的阿比梅渠化河道则经过整治工程而恢复了河岸带生态系统。住宅建筑前和公园边界都设置了带状共享花园，由居民共同栽植蔬菜，它们形成公园和住宅街区之间生机盎然的过渡地带，也因此而减少了必须由市政府维护的绿化空间面积。

每个街坊都在城市更新组织中获得整治，旧建筑的整修与新住宅的建设皆环绕着一个中央花园而进行。近日拆除的大型条状建筑楼长达250米，这项拆除工作使附近街区得以通过公园和马孔市中心"重新产生连结"，而这个崭新的城市景观则让每个人都能够徜徉其中，并加以利用。此规划工作通过地块切割以及边界与实用空间的设计，为街区重新建立了结构性与社会性的紧密连结。

此方案荣获2014年景观优胜奖中的街区整治类别银牌奖。

Residential blocks, a tree-lined trail and shared gardens

Previously isolated and cut-off, the Marbé Neighbourhood was the object of major work of urban renewal for ten years. The apartment buildings and towers seemed plopped down on large parking lots. And yet, this large area has benefited from great advantages because of its location and the available space. In Situ was retained to install the park and one part of the roads and city blocks.

This urban project is based on the reworking of land ownership between private and public spaces, and on the new grid of transversal streets that redrew residential blocks. The demolition of certain apartment towers made it possible to create an ample visual opening and to develop a long park promenade that runs across the area and is used for playground, sport fields and rain gardens. In the middle of the park, a canalized river, the River of the Abyss, underwent a restoration of its banks. At the foot of the residences, and at the edge of the park, strips of communal gardens are cultivated by the inhabitants. These parcels of nourishing vegetable gardens, active filters between the public park and the residential blocks, also reduce the surface area of green space maintained by the city.

Each block has undergone an urban transformation involving the rehabilitation of the buildings and the construction of new residences around a garden. The recent demolition of a very large, 250-meter long apartment building has allowed these quarters to reconnect to Mâcon's downtown, through the park. It is a new urban landscape explored and endowed by all. This work, focused on parcel division, borders and uses has made it possible for the neighbourhood to establish strong links, both structural and social.

This project received the silver Landscape Victoryies 2014, in the Neighbourhood Development category.

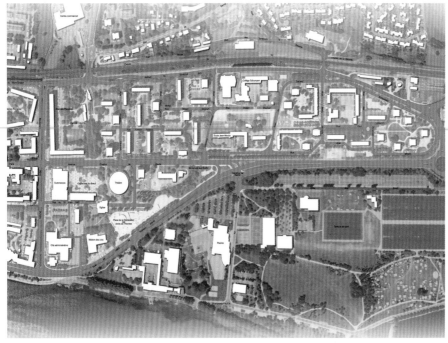

左上图：阿比梅，整治前，一条被嵌围住的溪流；整治后，可吸收洪泛的阶梯式绿化空间 / 右上图：总平面图
下图：散步道沿途的各种场景——树荫下的草地、樟子松……

Left: The Abyss. Before, a cut-off stream; after, floodable planted terraces / Right: Master plan
Bottom: The sequences of the promenade: shaded meadow, dotted with Scots pine...

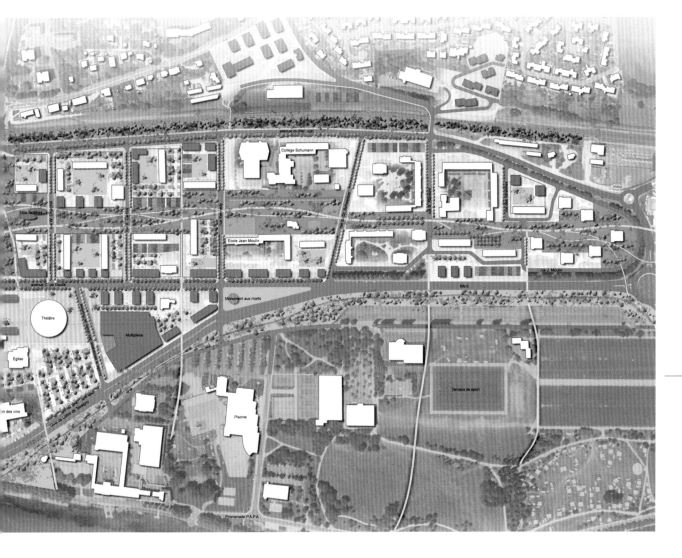

散步道公园,连接各种用途的整合性空间:菜园、运动场、游戏场、街区穿越道、住宅街坊通道……

The park promenade, connecting different functions: vegetable gardens, stadium, playgrounds, route across the neighbourhood, access to residential blocks...

上图：位于住宅脚下的共享花园
下图：位于公园边缘的带状花园形成一道"植物过渡空间"
Bas : Vue depuis le parc de la bande jardinée, une « transition cultivée »

Top: View of the communal gardens in front of the residences
Bottom: View from the park of the gardened strip, a "cultivated transition"

法国 巴黎 / 2007-2014

Rosa Luxemburg garden
罗莎·卢森堡花园

既开放又遮蔽的铁道公共空间

罗莎·卢森堡花园在巴黎帕若尔市集大厅的遮护下，沿着巴黎东站铁道在基地上从北向南延伸。这个带状铁道花园创造出一条延续不断的路径：从市集大厅的遮顶花园到邻接街区道路的露天花园。

露天花园从弗朗索瓦兹-艾莱娜·朱达整治的帕若尔市集大厅（青年旅馆、服务设施、商店与市立图书馆）北方展开。层层排列的平台上分置了阶梯座位、儿童游戏场和草地。一株株欧洲赤松高低有致地缀点着每一层平台，而主要路径边缘则种了一整排梣树。这片长坡形成一条慢慢滑进市集大厅和图书馆的"铁道"。

遮顶花园位于南边，阴凉花园和白色花园在巨大的钢铁棚架下开展，此顶棚经过转化改造俨然成为太阳能光伏发电中心。人行步道重现过去铁道的痕迹，沿路设置了多年生植物与地被植物园圃以及一些纵向水池。这些带状植物园圃利用储存在水池中的屋顶回收雨水进行灌溉，而这些水池自身也成为水生花园。地被植物、蕨类、小灌木、禾本科植物、藤本植物和丛生灌木在这个大型工业结构的屋顶下形成一种林内环境与氛围。这是一个邻接巴黎东站的植物绿洲，静谧而安详。两个共享花园沿着这些平行排列的绿化园区的边缘而延伸，拥抱着视野开放的巴黎东站及簇集密布的铁道，而散步至此的人们还不时能看到一列列火车的彩色车厢从眼前经过。

A public railroad space, both opened and covered,

In Paris, all along the railroad tracks of the Gare de l'Est and inside the Halle Pajol, the Rosa Luxemburg Garden stretches from north to south. The long train track outlines a continuous route that encompasses a covered garden under the Halle and an open garden in contact with the streets of the neighbourhood.

The open garden develops to the north of the Halle, restored by Françoise-Hélène Jourda (youth hostel, services, businesses and municipal library). Tiered terraces host rows of seats, grassy surfaces and a play area. A grove of Scots pine punctuate each of these landings while a border of ash trees accompanies the principal path. This long walkway mimics a "railroad" that slides smoothly under the Halle and the library.

To the south, the covered garden, shade garden and white garden, unfold under the vast metallic framework of the Halle, transformed into a photovoltaic power plant. The alleys, flanked with perennial flowerbeds and ground cover as well as longitudinal ponds, follow the traces of the old railroad tracks. The planted areas are watered by rain collected on the rooftop and stored in pools that become aquatic gardens. Ground cover plants, ferns, bushes, grasses, climbing plants and shrubs form an environment and an atmosphere of forest undergrowth within the shade of the vast industrial structure. It is a calm and peaceful oasis of vegetation that abuts the Gare de l'Est. Two parcels of shared gardens spread out on the edge of this landscape like theatre wings. These two foregrounds open on the unobstructed horizon of the Gare de l'Est and on the band of railroad tracks enlivened by the coloured trains that amblers watch as they pass by.

上图：帕若尔市集大厅整治基地全景
中图：遮顶花园整治前后实景
下图：露天花园整治前后实景

Top: General view of the site of the Halle Pajol
Centre: Before and after views of the covered garden
Bottom: Before and after views of the open-sky garden

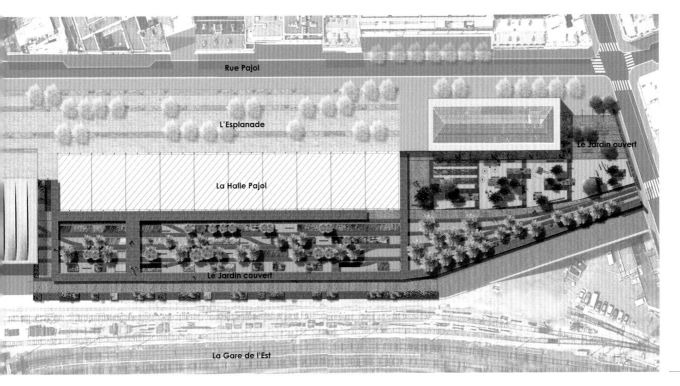

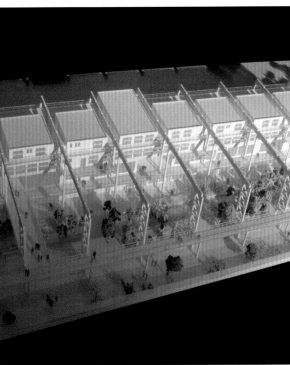
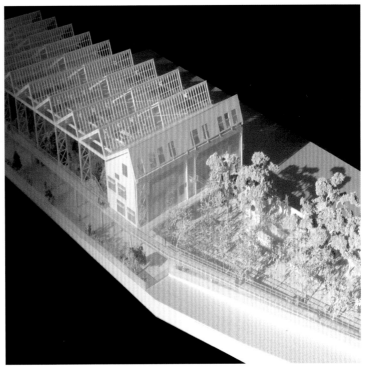

上图：总平面图（竞赛方案）
下图：方案模型

Top: Competition master plan
Bottom: Scale model

在市集大厅钢铁棚架的庇护下的遮顶花园　　　　　　　　　　　　View of the covered garden, under the metallic frame of the Halle

水生花园和带状植物园圃　　　　　　　　　　　　　　　　　　The aquatic gardens and the planted strips

左页：斜坡步道和露天花园、使用屋顶雨水浇灌的水生花园、露天花园里的儿童游戏场平台
右页：平行带状配置的花园空间，反映出基地的"铁道"特色

Left: View of the pedestrian ramp and the open gardens, the basins fed by water collected on the roofs, the playgrounds of the open garden
Right: Composition of corridors and the "remains" of railroad tracks

59

60

> "融入景观，便是成为广阔土地的一部分，
> 在此，一切事物之间的互动关系构成了庞大有力的生态系统。"
> 米歇尔·寇拉儒
>
> *"To be in the landscape is to be a part of a dimension
> where the system of interrelations is extremely powerful"*
> *Michel Corajoud*

places / environments
场所与环境

生态概念让人们逐渐意识到每个场所都是充满生机与活力的环境，都趋向一种本质上不稳定的平衡状态。每个环境都一直在转变，永远和诸多元素产生互动关系，例如空气、水、土壤、阳光、风、动植物，以及越来越城市化并使用着这些元素的人类。因此，大自然也越来越具有城市性格。这些生机勃发的环境彼此共处、不断地相互调整，没有停滞不变的事物，一切都在动，都在成长，都在逝去，在景观中彼此取代、移转……这个觉醒意识促发了全新的项目实现方式：启动演变进程；从生气勃勃的有利因素采取行动，而非着重无生命的物件；栽种日益成长茁壮的植物，以此来为城市地面与城市小品逐日毁损而失去光彩的场所带来活力；改善基地的土壤质量；准备、修复、耕翻、播种，并保存雨水以便灌溉植物。同时，也必须预先思考维护与保养的问题，也就是预测未来的发展，因为所有景观的主要原料都同时涉及生物、空间与时间……

Ecology has made it possible for us to become gradually aware that each place forms a living and dynamic environment that tends towards a naturally unstable balance. Each environment is in constant movement, in perpetual interaction with elements such as air, water, earth, sun, wind, plants, animals and humans, more and more urban, that use them. In fact, nature is also urban. These dynamic environments mingle with one another and alter one another. Nothing is fixed, everything changes, grows, dies, is replaced and moves in a landscape... This awareness leads to new ways of implementing projects: inspiring processes, putting in movement, acting on dynamic levers rather than on inert objects, hosting the vegetation that grows and beautifies there where the most sophisticated pavement and furnishings fade and tarnish, improving the quality of the earth in place, preparing, repairing, labouring, seeding and retaining rainwater so they can nourish the vegetation. But also, one begins to plan for the maintenance and upkeep of the space, that is, to anticipate what is needed because the principal raw materials of all landscape are living things, space and time all at once...

法国 圣马丁代尔 / 2005-2007

Lucie Aubrac place
露西·欧布拉克广场

如群岛般分布的雨水花园

位于格勒诺布尔市附近的圣马丁代尔镇在建设轻轨电车之际展开了一个整治此交叉路口的设计竞赛,以在中学旁边创造一个城市广场。尹西图事务所的方案企图在这片支离破碎的城市肌理中创造一个规整的长方形空间。然而,交叉路口的扩大以及建设计划的分段实施导致最初的设计构思被迫推翻,而改为施行另一个全然不同的方案。

伊泽尔省平原一带保留许多过去农地、池塘和排水沟渠的旧迹,并遍布着白柳和杨树。这些树丛吸收土壤里过剩的水分,塑造了城市的自然景观。尹西图事务所的方案对此农田组织进行了崭新的诠释,分别在不同的广场地块设置"花园岛屿"。这些尺寸不一、形似卵石的"雨水花园"根据远眺阿尔卑斯山脉顶峰的视角而进行配置,其规模大小则取决于需要吸收渗透的水量。这些略低于地面高度的花园里种植了大量的植物,周边以柳条编织矮篱来圈围,建立了一个个具有湿润环境的生态系统,也成为城市中独特的动植物天地。

这些花园岛屿的配置方式腾出了散步空间和一些小广场,以具有渗水性的稳定土铺设,并且靠着雨水花园设置了几条长椅凳。事实上,这个非典型公共空间的维护年费非常低廉。此花园广场展现了大自然也可具有城市特质的理念,而本方案也显示了一个事实:重新审视一项整治计划,以灵活的方式促使空间的演变发展,经常是明智之举,以便落实项目中最精髓也最有意义的部分。

An archipelago of rain gardens

Near Grenoble, the competition for the development of this intersection launched during the construction of the tram line had as its goal the creation of an urban town square at the edge of the secondary school. In Situ's design attempted to cut the regular figure of a long rectangle in this fragmented tissue. But the enlargement of the intersection, the phasing of constructions, led to a reformulation of the initial question and thus to the implementation of a completely different project.

The area around the Isère plain preserves the traces of the old agricultural land division, with pools and drainage ditches accompanied by large white willows and poplars. These clusters of trees suck up excess water and fashion the landscape of the city. The proposal to reinterpret the management of these agricultural lands consisted in installing an archipelago of garden islands on the different parvises of the surrounding buildings. These "rain gardens" take the irregular form of different-sized stones, positioned in order to take advantage of lookout points of the Alpine peaks and sized according to the quantity of water that infiltrates the area. These slightly sunken gardens are generously planted and ringed by a simple fence of braided willow. They host an entire ecosystem of wetlands, an unexpected flora and fauna in the city.

The arrangement of the garden-islands provides space for a strolling area and for several parvises coated with a stabilized permeable surfacing. A few benches abut the rain gardens. In practice, the cost of annual maintenance of this atypical public space proves to be very modest. This garden-square demonstrates the idea that nature can also be urban. It also illustrates the fact that it is often relevant to re-examine a development plan, to know how to make it evolve with flexibility in order to produce only nothing but the essential. And that only makes sense.

上排图：柳条编织围篱形成了"花园岛屿"的边界 / 花园的照明设施
中排图：雨水花园 / 鸢尾与柳树
下排图：一个"花园岛屿"的入口 / 可吸收洪泛的雨水花园 / 携带着信息的苗木支柱

Top: The gates of braided willow installed during the planting / Lights on in the garden
Centre: The rain garden, today / Iris and willows
Bottom: The gate of a "pill garden" / The rain garden, today / A marked garden stake

上图：雨水花园景观，建造后的第一个春天
下图：雨水花园景观，五年之后……

Top: View of the rain gardens, first spring
Bottom: View of the rain gardens, 5 years later...

法国 里昂 / 1999-2005

fort Saint-Jean
圣让堡

环堡巡防步道和花园观景台

圣让堡矗立在山石嶙峋的雄伟山口上，位于穿越里昂红十字山坡的索恩河狭道左岸，过去是里昂市北方防御边界的军事堡垒，建于16世纪，经过重大变更后，从此成为国立国家金库公务员培养学院。

这个方案设计强化了本基地的优势，突显出俯瞰里昂市的视野角度。阅兵广场以石灰岩作为铺地材料，广场中央的三棵榆树为堡垒主入口提供了框景。皮埃尔·维尔帕斯联合建筑师事务所设计的新建筑物宛如"景观建筑"，采用了军事堡垒的平台、斜堤、墙体等建筑语汇，与环境产生协调的关系。新设餐厅的屋顶阳台可观看市景，从地中海式花园可远眺索恩河狭道和福维耶山岭的美景，花园由几列低矮的开花长青灌木丛和多年生芳香植物共同组成。

"巡防步道"成为主要的散步路径，呈现出各种不同角度的视野，并将不同平台串连起来。透过基地本身和日照方位之间的关系，堡垒中的每一个场所发展出一系列对比强烈的植物情调：硬铺地广场明亮而日照充足；棱堡花园既存的法国梧桐枝繁叶茂，形成林地氛围；角面堡处可见到松树和紫荆树；图书馆花园的无患子树枝叶扶疏，禾本科植物茂盛繁密有如地毯；高悬的格栅过道上攀爬着紫藤；多风且阳光曝晒的低处平台栽种无花果树和蜀葵；而草地上则盛开田野花朵。这些花园观景台巧妙地为严谨、呈几何形状的军事要塞以及较为奔放、任植物自由生长的自然空间做了连结。

A walk along the walls and through gardens with views

Fort Saint-Jean perches on an imposing rocky spur, on the left bank of the gorge of the Saône River that forms the Croix Rousse hillside in Lyon. This former military fortress, erected in the 16th century to defend the northern limit of the city, has undergone major changes, and currently is home to the École nationale du Trésor public (National School of the Public Treasury).

The finished project improves the quality of the site and the vantage points that it offers of the city. At the centre of the parade grounds, paved with limestone, three elms frame the principal entry. The new constructions of Pierre Vurpas & Associates resemble "landscape-buildings". They integrate smoothly into the fortress, employing the vocabulary of terraces, fortifications and walls. On the roof of the new restaurant, laid out as a balcony over the city, a Mediterranean garden opens to views of the Saône and Fourvières. It is composed of several lines of low, flowering evergreen shrubs and scented perennials.

The central theme of the circuit, the "wall-walk" spotlights different look-out points and connects the different terraces. In each of these places, the relationship between site and its sun exposure allows the development of contrasting mineral and vegetal environments: the central place is dominated by stone, luminous and sunny; the fortress garden has a woodland environment beneath the foliage of the existing plane-trees; the redoubt, a defensive structure, features pines and Judas trees; the library gardens use soapberry bushes and a carpet of grasses; a high footbridge functions as a trellis where the wisteria clings; the low terraces, subject to the wind and sun, feature rustic flowered meadows, fig trees and hollyhocks. These look-out gardens marry the geometry and military rigour of the fortifications with spaces of a wilder nature where vegetation can freely spread.

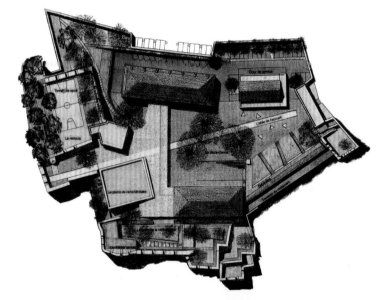

左页上图：从福维耶丘陵看向堡垒 / 总平面图
右页上图：阅兵广场和地中海式花园
下排联图：昔日的环堡巡防步道，如今成为一条连接着花园观景台的散步道

Left: View of the fort from Fourvière Hill / Master plan
Right: View of Arms Place and the Mediterranean Garden
Bottom: Views of the wall walk, a promenade linking the viewpoint gardens

位于餐厅屋顶平台上的地中海式花园,黄杨、岩蔷薇、薰衣草、百里香和迷迭香,这些植物特别适合干燥、日照充足的平台式基地

The Mediterranean garden installed on the roof of the restaurant, boxwood, rockrose, lavender, thyme and rosemary, species adapted for a dry and sunny terrace site.

花园平台：充满田野气息的各种植物，随着每个地块的不同日照朝向而组织，呈现对比鲜明的氛围；过道格栅则上攀爬着紫藤

The garden terraces: a rustic vegetal palette that forms very diverse ambiences according to the sun exposure; the trellis-passageway planted with wisteria

法国 圣沙蒙 / 2010-2012

the changeable park of the Aciéries
炼钢厂演化公园

结合工业和大自然的城市公园

吉阿镇位于吉耶河谷地，从工业革命初始便和圣沙蒙市的发展历史息息相关。这个基地的兵工厂关闭以后，留下了一大片空洞、贫瘠又孤立的闲置空间。尹西图景观事务所和贝尔纳·巴里建筑师事务所共同进行了这个城市设计项目，将重心放在这个机会难得的空地上：修复工业厂棚并迎接全新的产业活动，以便更新基地面貌、沿着基地边界规划一个长形开放的散步公园。

为了尽快实施这个转变计划，尹西图事务所在六个月内以相当有限的预算完成了第一部分的公园整治：方案通过围墙的拆除而得以打开视野，并重建和城市的关系。炼钢厂公园沿着一条供行人和自行车使用的主要路径来组织空间。停车场和建筑物拆除后的空地上种植了一片花草地，其间点缀着繁茂的树丛，以此分割草地空间，并界定一些较为隐秘的场地、儿童游戏场、户外舞台、溜冰场。基于对地面污染的考量，事务所采用了适合贫瘠土地的整治方式，设置了栽种桦木与禾本科植物的植物小岛，为公园提供一层厚厚的表层土壤。过去工业活动的各种遗迹分别被展现出来：古老的淋浴设备、墙面和旧脸盆……原地面也被保存下来，转化为入口广场或走道。埃勒克斯恩可公司设计的标识系统揭示了基地的某些特殊元素，例如大烟囱、管道化的吉耶河旧迹、某工坊门槛等等。

这个具有演化性格的公园，将逐渐扩大，并通过新设备的设置而更加完备。然而，这个景观当前已经以极少的花费恢复了活力，再度成为当地居民可享用的空间，原来的工业荒地化身为公园和散步道，成为一个开放而生气勃勃的生活场所。

A Park that unites industry and nature to the city

In the Giers valley, the closing of the Giat weapons factories left vacant a vast space, gaping, sterile and isolated. The foundation of an urban project led by the Atelier Bernard Paris, this open space is an extraordinary opportunity: the rehabilitation of industrial halls and the hosting of new activities permitting the renewal of the site, on the edge of a long, open, park-promenade.

In order to jump-start a rapid transformation, In Situ realized the first part of the park in six months and with a very small budget: the demolition of the enclosing walls permitted the re-establishment of views of and connections with the city. The Changeable Park of the Aciéries is organized all along the principal pedestrian and cycling path. A flowered meadow was seeded where parking structures and demolished buildings stood. It was dotted with wooded groves that broke up the expanse and defined more intimate spaces, a playground, an outdoor theatre, a skating path. Taking into account the ground pollution, the topsoil that was brought in was distributed in thick layers in the form of planted islands of birch and grasses, adapted to the poor soils. Different industrial remains have been incorporated in to the design: factory workers' showers, walls, an old basin… The pavements have been preserved and transformed into a parvis and walkways. The signage of Alexandco maintains certain unique elements, such as the large chimney stack, the outline of the Giers River, canalized and piped, the threshold of a workshop, etc.

This park, design to evolve progressively, will be enlarged and completed by new facilities. But the landscape was already revived at very low cost and reinvested by the residents: this industrial wasteland has become a park and a promenade, a place of open and fertile life.

公园细部　　　　　　　　　　　Details of the park

一个公园，多种场所：草地、游戏场，以及特别设计的标识系统

A park, several places: the meadow, the playground, scenographic objects

法国 圣克里斯托尔 / 2008-2013
the ampelographic gardens of Viavino
维阿维诺葡萄品种栽培园

朗格多克地区的葡萄酒文化和景观

吕内勒城市共同体决定在圣克里斯托尔建造维阿维诺中心，这个与众不同的项目以展现葡萄酒酿造工艺为主。尹西图事务所和菲利普·马德克事务所联手合作，为这个地域规划项目进行设计，以既兼具当代感又符合农业需求的协调方式，将建筑物体量和自然景观结合在一起。七栋建筑组成了类似开放性的村落，融入断断续续的村庄肌理中，其建筑体量也与整体环境达成和谐。这个景观方案延续了地块的组织结构，并串连不同功能的建构物：酒窖、工坊、门市展厅、酒吧柜台、圆形剧场、内院、餐厅。

北方边界的松树和冬青栎郁郁葱葱，构成这个葡萄工艺观光中心建筑物的屏障，挡风之余，同时借此与现存住宅之间保持适当的视线距离。在东边，入口广场延伸至建筑物之间，直到与朴树成荫的平台衔接。平台下，栽培各种葡萄品种的花园朝南方展开，直到大型暴雨池的边上。从梯田式的地块和大型棚架可看到各式各样品种的葡萄苗栽培：歌海娜、慕合怀特、丹魄、佳丽酿、神索、西拉、梅洛、赤霞珠……此外，园区中还设置了许多附加用途的场地，如：芳香植物花园、木制设备游戏场、露天剧场、橄榄树林、玫瑰园。

本项目显示了某种现代乡村的形式：重视生态、简朴与共享的环境。此方案设计汇聚了公共与私人范畴的资源，才得以创造出这个创新的基地。这是一个全新的生活场所，既具有凝聚力又拥有实质意义，此环境以具有教育效果的方式将适度而可持久发展的葡萄栽培呈现出来，为大自然与文化之间建立了良好联系。

Landscape and wine growing in Languedoc

The urban community of Le Pays de Lunel commissioned, in Saint-Christol, the construction of Viavino, an atypical project of this cenotouristic hub. In close collaboration with Atelier Philippe Madec, In Situ conceived a large-scale project that blended constructed volumes with the landscape according to a cohesive programme that is both contemporary and agricultural. Seven buildings compose a sort of open hamlet that join the discontinuous tissue of the village without creating ruptures by means of their volumetric architecture. The landscape extends the organization of the agricultural parcels and links the different constructed entities: wine cellar, workshops, market hall, bar, amphitheatre, patio, restaurant.

To the north, a border area lined with pines and evergreen oaks abuts the buildings of the wine tourism centre, to protect it from the wind while also maintaining the legal minimum distance in relation to the existing residences. To the east, the entry parvis extends between the buildings to the terraces shaded by hackberries. Further down, the ampelographic gardens spread out to the south up to the large retention basin. Terraced parcels and large trellises display the cultivation of different varieties of grapes: grenache, mourvédre, tempranillo, carignan, cinsault, syrah, merlot, cabernet sauvignon... A garden of aromatic plants, a playground with wooden equipment, an open-air theatre, an olive grove and a rose garden create different areas of complementary uses.

This project illustrates a certain form of modernity, rural, ecological, frugal and shared. The pooling of public and private resources fuelled the rebirth of this innovative site. A new living space makes connections and makes sense, an environment that unites nature and agriculture, through the educational expression of sustainable and well-reasoned grape-growing.

上排图：露天剧场 / 葡萄园和围栅
左栏图：儿童游戏场
右栏图：玫瑰园圃和绿色剧场阶梯 / 中庭与围栅 / 干石矮墙

Top: Open-air theatre / Area with vines and arbour
Left: Play area
Right: Rose bushes next to planted amphitheatre seats / Patio and arbour / Low dry stone walls

上图：基地全景
下图：从入口广场看到的景致

Top: View of the entire project
Bottom: View from the entry way

法国 圣普利斯特 / 2013-2014

Nelson Mandela park
纳尔逊·曼德拉公园

一个融合自然、农地与城市特质的公园

圣普利斯特镇位于大里昂城乡地区东边广大平原的中心地，历经了近五十年来混乱的城市发展过程，在城镇中心和贝莱尔社区之间，以及墓园、小学、中学与高中后面一带，却意外地出现了一个建筑体量的"缺牙"，其轮廓随意而不规整。这块被人遗忘的12公顷土地却又像是一个被特别保存的自由空间，一个能让人喘息、极目远眺的绝佳场所。如此广阔的空地，只等着让人为它重启生机。

设计竞赛要求参赛者规划一座城市公园。尹西图事务所提出的方案对于形成此基地特征与地理形态的三种对比强烈的生态环境加以强化：高原、山坡地和湿地平原。雨水收集的考量界定了排水草沟的路径安排，草沟中的雨水最终将注入池塘和暴雨池。公园的四个入口通过一些主要通道互相连接，而这些排水草沟便沿着主要通道而设置。三种生态环境依各自的自然特性提供了不同用途的场所：
- 位于高原上的"欢乐草地"，其边界地带设置了集体经营果园与共享花园；
- 呈带状且种植绿树的"梯形地"，其中设置了旱地花园与运动场；
- 点缀着松树丛的"高草原"，供人在此进行野餐，享有面对池塘与里昂附近山岭的景观；
- 具有游戏设施的"趣味平原"，能接收过多的雨水。

这些生态环境都是充满生机的场所。为了建立一个"自然的城市公园"，也就是一个公园生态系统，方案预先规划了一种简约而经济的空间管理。然而上一轮市政选举的结果过造成了市长的更换，这个获得设计竞赛首奖的方案也就此停止，设计构思的任务也戛然终止。这块土地的未来终将如何？房地产开发、停车场或者成为墓园的延伸？

A natural, agricultural and urban park all in one

In Saint-Priest, in the middle of the vast plains of the greater Lyon area, the chaotic urbanization of the last fifty years has left a curious large hollow carved with irregular contours, between the centre of town and the Bel Air Neighbourhood, and behind the cemetery and the primary, middle and high schools. This forgotten territory is an astonishing preserved free space, an open breath on the horizon. The immense wild area of 12 hectares constitutes a generous empty space that begs to be revived.

The competition framework required a design for an urban park. In Situ proposed to highlight and strengthen the identity and geography of the site, made up of three contrasting environments: the plateau, the hillside and the wet plains. Rainwater collection delineates the route of drainage ditches that feed a pond as well as storm water detention basins. The ditches make use of the lay-out of the principal paths of the park that connect the four entrances.

These three environments by their very nature host different uses:
- the festive meadow on the plateau, a border of collective orchards and community gardens;
- the Trapezoid, a planted strip filled with dry gardens and sports fields;
- the high meadows punctuated with pine groves, available for picnics, with a view on the pond and the Monts du Lyonnais;
- the playing field, equipped for outdoor activities, which can also collect run-off rainwater.

All of these environments are living places. In Situ has planned for a low-maintenance and cost-effective management of these spaces in order to create a "natural urban park", a park ecosystem. Following a mayoral change after the last municipal election, the prize-winning project was suddenly halted and the project manager of the team had his contract terminated. What will this area become in the future: subdivisions, parking lots or an extension of the cemetery?

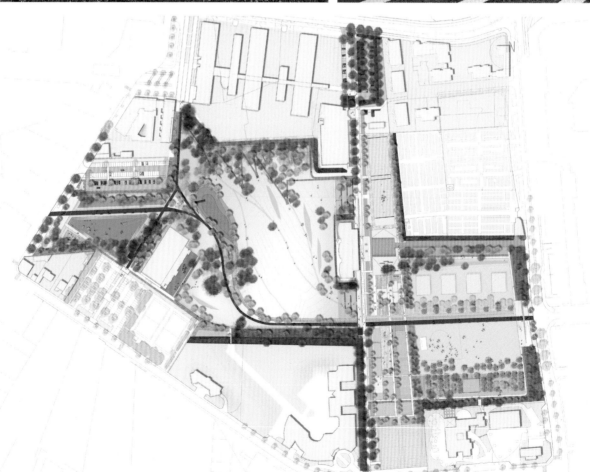

上排图：基地现况 / 游戏平原透视图
下排图：总平面图 / "梯形地"和位于坡地上的池塘

Top: Actual state of the site / Perspective view of the game fields
Bottom: Master Plan / Block diagram, the "Trapeze", and the pond on the slope

"有时候，当城市肌理松散一点、天空下降一点，
风景便会进入城市：河流的穿越便是最生动的例子。"
米歇尔·寇拉儒

*"Sometimes, the landscape enters into the city, when the urban mesh relaxes
and the sky descends: the passing of the river is the most fertile example of this"*
Michel Corajoud

岸 地 与 河 流
banks / rivers

自然景观是由共用的线条所构成的，谁能分辨河畔与海岸的蜿蜒曲线是来自河流、海洋还是陆地的线条？这场线条的拉扯战仍在持续进行中，时而顺畅、刚劲有力，时而激动猛烈，由能量充沛、生气勃勃的各种岸边生态环境所左右着：在河水上涨、潮汐涨落和枯水期等的有节律侵袭过程中，河岸林地、潮间带、泛洪漫滩和坡堤的组合存在着脆弱的平衡关系，其边界线条既是变化无常又是模糊难辨的。在河流与岸地、江川与坡堤、滩地、海洋、海岸之间，这片界限不明的"滩段"上演了许多动荡且动人的情景。身处水边，便是处在一个奇特的中间地带，一切都在变动，一切都在转化。河流为岸边茂盛的植物提供水分，而植物则减缓水流流速。江川、坡堤、海岸等等空间，都是"和生物共同创作"的最佳场所。整治岸地，意味着采用环境本身所具有的各种生物活力来伴随和引导它们的演变。

Landscapes are made of watersheds. Does the river, the sea or the earth draw the sinuous lines of the banks and the shores? This continuous negotiation, at once fluid, vigorous and violent, makes riparian environments the most energetic and dynamic in existence: in rhythm with the tides, the highs and lows, the riverside vegetation, the beaches, the floodable strands and the embankments form worlds in precarious balance, fluxes that are both fluid and fitful. Between the river and the banks, the river and the embankments, the strands, the ocean, the coasts, this uncertain "beach" creates shifting and stirring conditions. To find oneself on the edge of the river is to be in a unique in-between where everything moves and transforms. The river feeds the vigorous vegetation of its banks which in turn slows its current. Rivers, embankments, coasts are choice places to work "with the living". Developing banks consists of accompanying and guiding their movements, employing all the dynamics at work.

马提尼克 法兰西堡 / 2002-2015

the large Savannah
大 拉 萨 瓦 公 园

两条林荫道、一个拥抱海洋的公园和甲板平台

拉萨瓦区这个历史性空间位于碉堡下方，占地7公顷，是法兰西堡市中心的标志地点。这个过去防卫用的缓坡地慢慢转化为阅兵广场，继而变成公园和停车场。与此同时，市中心的居民大量外移，拉萨瓦区也变得支离破碎，杂树丛生，在无人管理的情况下日渐式微。

尹西图事务所的方案于2002年从设计竞赛中脱颖而出，直至今日虽然仍未全面落实，却为拉萨瓦区这个重要地点重新注入了活力，并重新建立该区和弗拉蒙海湾之间的关系。两条宽敞的林荫道界定了中央区域，并延伸至加勒比海边。地面彩色铺地的网格犹如马德拉斯花格布，延续了这个殖民地城市的方格状城市肌理。图书广场和加勒比海大道面对着舍尔谢图书馆铺展开来，自由大道上的桃花心木树荫下则设立了许多售物亭。靠海滨的快车道被转化成城市大道，而大型甲板平台和层层阶梯则将滨海散步道延伸至法国海滩，让人们又再度能够在城市空间里戏水游泳。

点缀着高耸大王椰树的中央大草坪重新成为完整的空间，并拥有面海的视野。曲折蜿蜒的散步道连接了七座花园岛屿：游戏岛屿、记忆岛屿等等，展示出一个个热带植物的迷你天地，显现不同的氛围与用途。利奥波德·塞达尔·桑戈尔过去送给艾梅·塞泽尔的种子长成的猴面包树被移植到拉萨瓦区的中心位置。许多活动都在这个活力充沛的公共空间里举行，如小艇竞赛和著名的狂欢节。

此方案曾荣获多重奖项，包括2014年城市艺术奖所颁发的市长奖（罗伯特·乌泽勒研讨会）、2007年法国城市入口奖，以及2002年法国电力集团设计竞赛首奖。

Two promenades, a park and a deck open to the sea

The Savannah is a historic and emblematic space of 7 hectares in the centre of Fort-de-France, at the foot of the fort. This ancient defensive fortification has been little by little transformed into a parade ground and then a parking lot. During the same time, the heart of the city experienced an exodus of its inhabitants and the Savannah, parcelled up and buried under the trees, became run-down.

The In Situ project, a 2002 competition winner as yet incomplete, made the renewal of this major place possible, re-establishing the link that unites the Savannah to the Bay of Flamands. Two large promenades frame the central expanse and reach all the way to the Caribbean Sea. The weave of the coloured pavement coating extends the square-grid urban fabric of the colonial city, like a Madras textile pattern. The Parvis of Books and the Boulevard of the Caraïbes unfold across from the Schœlcher Library. The Liberty promenade is dotted with many kiosks in the shade of mahogany trees. The beachfront highway has been transformed into an urban boulevard. A large deck and stairs extend the seaside promenade up to the beach of the Française, where it is once again possible to go swimming in the middle of the city.

The vast central lawn, punctuated with tall royal palm trees, has found its unity and its sea view. A serpentine promenade connects seven garden-islands, so many microcosms that highlight the tropical flora, its ambiences and different uses: Islands of Games, of Memory, etc. The baobab, grown from seeds given by Léopold Sédar Senghor to Aimé Césaire, was transplanted to the heart itself of the Savannah. This active public space hosts numerous shows, skiff races and the famous Carnaval.

This project received the Mayor's Prize 2014 for Urban Art (Robert Auzelle), the Prize for City Entrances 2007 and was the winner of the EDF Competition 2002.

左上图：整治前后的海岸甲板实景
右上图：总平面图
下排连图：方案的不同场景——沿海散步道、拉萨瓦大草地和堤堰

Left: Before and after views of the seaside, easement of the deck
Right: Master Plan
Bottom: Sequences of the project: the sea-front promenade, the large meadow of the Savannah and the jetty

场景拼图：沿海甲板、弗拉蒙海湾、堰堤、草地、植物、游戏场、散步道上的售物亭、繁茂的植物、花草近景

Mosaic: the large sea-side deck, the Bay of Flamands, the jetty, the meadow, details of the plantings, the playgrounds, the kiosks of the promenade, the luxuriant vegetation

甲板、散步道和大草地　　　　　　　　　　　　The deck, the promenade and the large meadow

法国 里昂 / 2003-2008

the banks of the Rhône
罗纳河岸

河港、公园与河畔散步道

罗纳河作为里昂城市边界为时甚久，在过去，吉约蒂耶尔桥是唯一穿河而过的桥梁，而河水上涨时波涛汹涌，水势惊人，限制了河左岸的城市化发展。到了19世纪末期，石砌护坡的建设使城市得以越过河岸向东边扩展。直到1960年代，这些废弃的低码头变成了可容纳1,600个车位的大型停车场，罗纳河成为一条被车辆所占据的河流……然而，这个占地数十公顷、延绵5公里长的"中间地带"俨然是城市可利用空间的保留地，它既处城市边缘，又位于市中心。这个更新河岸用途的项目反复酝酿，并成为一项设计竞赛的主旨，于2003年由尹西图事务所夺标。此河岸改造的重新启动，历经研究设计、商议和工程实施，在短短五年的时间内完成。在低码头的石铺地面下，一个城市河滩等待着重新现身……

获选的设计初稿以基地卓越的环境特色和一些简单而柔和的主要线条为基础来发展设计构思，这些线条确保了这个沟地自然景观的连续性，也主导了介于金头公园和杰尔朗公园之间、沿着水流设置的慢行交通路线。在横向部分，方案依据散步道的不同宽度（从7米到70米不等）发展了多种用途。这块土地为城市里提供了绝佳的喘息空间，本方案的目标便在于加强基地的特质。从上游到下游，一系列沿着水岸设置的场景循序展开，形成长长的连续景致，并由十座桥梁谱出韵律节奏：这个线形基地的上游与下游处最是绿意盎然，而越往中央段落前进则越是采用硬质材料。上游河岸林地与支流水潭地的自然空间相得益彰。具有实用功能的河畔则停靠了一整列船屋，并有满布柳树和杨树的花园岛屿点缀其间。大草坪上尽是野餐的人们，而"板条散步道"则同时形成一条充满欢乐氛围的长条状露台，和游艇餐厅相衔接。

在河道弧线的中央位置，宽广如露天剧场的吉约蒂耶尔阶梯式平台面朝西边，放眼望去便是辽阔的水平线、罗纳河与其支流的景致。在水上活动中心处，防栅凸堤被扩大延展成临河看台，而在橡树与榆树绿荫遮护下的大学码头则专供游轮

A harbour, a park and a riparian promenade

For a long time, the Rhône River in Lyon was a boundary traversed by a single bridge, the Guillotière, and the extreme rises in water level limited the urbanization of the left bank. At the end of the 19th century, the construction of the dry stone wall enabled the village to spread out in the east, beyond the left bank. Then, in the 1960s, the part of the harbour closest to the water was transformed into an immense 1 600-place parking lot, a river of cars... However this "in-between" of some 10 hectares, 5 km in length, represented an astounding reservoir of available space, situated both on the edge and at the centre itself of Lyon. This restoration took place in the space of five short years of study, dialogue, and construction. Under the cobblestones of the ports, the beach awaited...

The winning design made use of the excellent quality of the site and of the simple, supple strong orientation lines that affirm the continuity of a natural channel and of a soft-means path that gently follow the water, between the Park de la Tête d'Or and Gerland Park. On either side, the promenade has been developed for multiple uses according to its different widths (from 7 to 70 meters). Along the river, several spaces follow in sequence and form a long tracking shot rhythmed by ten bridges: varying from the most vegetation on the path up- and downstream to a greater presence of stone in the midsections. The riparian woodlands upstream flourish around the meanders. The inhabited banks are lined with houseboats and punctuated with garden islands of willows and poplars. The large meadow is surrounded with picnickers while the "planks" form a long festive terrace with barge restaurants. The aim of this project is to strengthen the identity of the area, which is a magnificent breath of fresh air in the city.

At the centre of the arc of the river, the large amphitheatre of the terraces of the Guillotière opens to the western horizon, the Rhône and the tributaries of the river. To the right of the nautical centre, the loading dock is enlarged like a balcony over the river. Under a canopy of oaks and elms, the university harbour hosts cruise ships. Downstream, the waterfront road has given way to a fluvial botanical garden and

停泊。在下游,原本的滨河公路被改造成一条河边植物绿廊,恢复了河岸的自然景观。儿童游戏场、滑板区、滚球场和运动场点缀着各处闲置的地块。系船柱不仅具有泊船系缆的功能,也成为照明设备的元素。这条慢行交通路径由人行步道和自行车道互相交错所组成,呼应了罗纳河特殊的水线"编织"现象。河水涨潮的问题迫使设计必须避开所有障碍物,此一重大限制却促使方案创造出一个流畅、自由而开放的独特景观。

此项目的独特之处也在于它的"生产方式":其功能计划是依照每个场所的特质和使用潜力而建立的,并在早期阶段便开始进行具有建设性的协商,同时也和业主与使用管理部门预先为项目的后期管理与维护进行了规划。自此以后,还记得这块土地昨日仍旧充斥着汽车的人寥寥无几。如今,这些场所全然展现各自的特色:此河岸公共空间成为一道活泼丰富的城市线条,每天都有数千名使用者在此自由空间来来去去……

此方案荣获2008年城市整治锦标奖的特别佳作奖、2012年景观优胜奖,以及2012年《空间&时间》特别奖。

the river banks have been restored. Here and there, playgrounds, skating rinks, bowling greens and sports fields dot the available space. Boats are docked with the help of mooring dolphins that bring attention to this exceptional space. Situated in the river bed itself, the soft-means transport consists of pedestrian and biking trails that interlace, in imitation of the hydraulic phenomenon of tressage, or weaving, native to the Rhône River. The rises in the water level have necessitated clearing the river of all obstacles. This heavy constraint has made it possible to fashion a landscape that is singularly fluid, free and open...

The specific character of this project also holds true for its method of fabrication: the project was co-constructed according to the purpose of each place and its potential uses. Upstream, it was the subject of a constructive dialogue. Downstream, the management and maintenance were anticipated with the project manager owner and the uses management. By now, few people remember that this area just yesterday was still crowded with cars. Today, these spaces fully express their magnificence: a long, fertile muscle of public space, a space of freedom frequented each day by thousands of visitors.

Special Mention, Urban Development Trophy 2008 and Landscape Victories, Special Prize "Space and Time" 2012.

95

上排图：河岸空间的多样化用途和场景……散步道、花园岛屿、植物丛、自行车道、板条铺地平台、高堤岸和低堤岸之间的游戏场……

Top: Many uses at water's edge… / View of the promenade, sequence of the garden-islands; the different strata of the embankments: plantings, bike path, garden-islands, pedestrian promenade, planks for terraces… Playgrounds, slides between the low and high quays

下图：场景图——布雷蒂洛沼泽公园、上游河岸林地、花园岛屿

Bottom: Scenographic view, sequences: The Brétillod eco-garden, the river vegetation upstream, the garden-islands

上排图：上堤岸与下堤岸的交错 / 开放的大草地 / 散步道与露天咖啡座 / 吉约蒂耶尔阶梯式平台和浅水池 / 从散步道看阶梯式平台 / 浅水池

Top: Crossings, lower and upper quays / The open meadow / Promenade and terraces / The rows of the terraces of the Guillotière and the inlet / View of the terraced seats from the promenade / The inlet

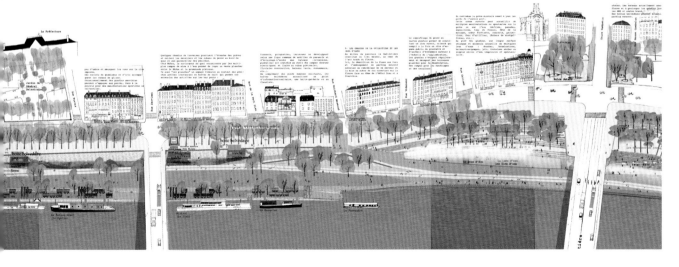

下图：场景图——罗纳河大草地、吉约蒂耶尔阶梯式平台

Bottom: Scenographic view, sequences: the large meadow of the Rhône, the terraces of the Guillotière

 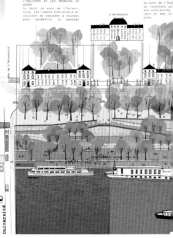

上排图：夏日戏水的场所 / 靠近水上活动中心的阶梯平台 / 滑板公园和运动场地 / 大学码头、休闲空间与游艇停靠处 / 正在防栅凸堤步散道上进行艺术创作的菲利普·法维耶 / 水上活动中心旁的防栅凸堤步散道

Top: Summer use, the water jets / View of the terraced seats on the side of the nautical centre, view of the beginning of the landing stage / View of the skate park ramps and the sports fields / The port of the University, space for relaxation and mooring of cruise boats / Artistic intervention of Philippe Favier / View of the landing stage from the nautical centre

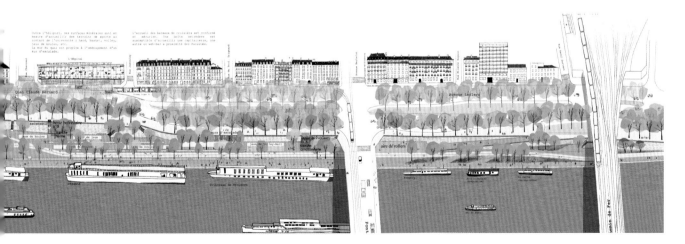

下图：场景图——水上活动中心处的防栅凸堤、大学港

Bottom: Scenographic view, sequences: the landing stage of the nautical centre, the port of the University

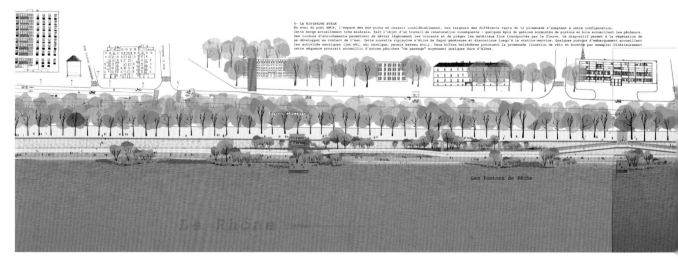

左栏图：健身区和滚球场
右栏图：散步道上的行人与自行车……
下图：场景图——下游河岸林地

Left: The area dedicated to fitness and the pétanque strips
Right: The promenade continues, pedestrians and bicycles...
Bottom: Scenographic view, sequences: the riparian vegetation downstream

左图和中上图：下游河岸林地、沿着散步道时而出现的浮桥与游戏场
中下图：散步道旁繁茂的大戟，远处可见罗纳河汇流处博物馆
右图：水岸边的钓鱼者与坐下小憩的散步者

Left: View of the riparian vegetation downstream, the pontoons and game platforms rhythm the path
Centre: View of the abundant plantings of euphorbia and the Museum of the Confluences in the background
Right: Walkers and fishermen at the water's edge

法国 索恩河畔罗歇塔耶 / 2011-2014

the promenade of the cafés
小 咖 啡 馆 散 步 道

在索恩河畔

罗歇塔耶基地位于里昂北方、索恩河左岸,全长两公里。这一带河畔内岸不受水流威胁,视野开阔,放眼可见金山矗立在远方,并形成一条悠长的凸曲线,点缀着沙岸与卵石滩。在20世纪初期,这里曾是最受里昂人喜爱的沙滩,小咖啡馆群聚遍布。这项用途虽然至今仍旧持续,汽车却占用了很多空间:省级公路切断了此地和村庄的联系,并且到处充斥着任意停放的车辆。

本方案的重点在于重现旧堤防痕迹,并在这条纤道上恢复具有连续性的宜人散步道。方案设计降低了汽车流量与污染,以便安排一条供行人与自行车使用的步道。有松动坍塌之虞的河岸以卓越非凡的植物技术加以强化稳固,而自然景观、村庄和城堡的横向视野也重被新建了起来。中央空间经过重新规划,成为供人休憩小睡的大草坪,草坪缓坡逐渐下降,一直延伸至河边。这个河畔草坪也可作为音乐会和各种不同露天活动的场地。起伏的地面上摆着一些躺椅,还设置了大桌子等着前来野餐的人们。

一系列不同的场景沿着这条河岸路线展开,犹如电影中的移动镜头一路显示的景致:纤道、儿童游戏场、砾石岸和草地、岛屿和水潭地、沙滩小径、小咖啡馆和钓鱼浮桥。这条路线沿途设有许多艺术装置,形成不同场所的标志:莎宾娜·郎和丹尼耶尔·鲍曼设计的平衡阶梯、善良男孩工作室的陨石、川俣正的树屋、福斯蒂诺的两面大镜子。每个作品都展现出对景观的独特诠释。这条艺术之路名为"河流电影",继续沿着索恩河畔的其他七个区段发展延伸。

The banks of the Saône

North of Lyon, the site of Rochetaillée stretches across 2 km on the left bank of the Saône River. This internal bank protected from currents, opens onto a view of the Monts d'Or and sketches a long, convex curve, punctuated by pebble beaches. In the beginning of the 20th century, it was Lyon's favorite working class beach, with its open-air cafés and dance halls. The tradition lasted for years, but cars have since confiscated lots of space: the departmental road severed connections with the village, and off-road parking was extended.

The major challenge of this project was to reveal the presence of the old dike and to re-establish the continuity of a gentle promenade on the haul road. Traffic circulation was reduced and calmed in order to preserve a path shared by pedestrians and cyclists. The weakened banks were shored up by a brilliant use of vegetation. Cross-sectional views of the landscape, the village and the chateau were re-established. The central space was reconfigured to form the large siesta meadow that gently descends to the river. This riparian meadow can also host concerts and various open-air shows. Lounge chairs are formed using grassy taluses and large tables await picnickers.

A progression of scenes unfolds along the banks and forms a sort of cinematic tracking shot: haul road, playground, beaches and meadows, meanders of the river, beach paths, open-air cafés and fishing piers. The route is marked with several artistic installations: the stairs to nowhere of Lang and Baumann, the meteorite of Gentil Garçon, the perched hut of Kawamata, and the two large mirrors of Faustino. Each of these works creates a unique perspective on the landscape. This global artistic route, called the "river movie", unfolds over seven other sequences of the Rives de Saône.

上排图：纤道与草地，整治前后实景 / 总平面图
下排连图：河岸散步道的不同场景

Top: Before and after view, the haul road, the meadow / Master plan
Bottom: Views of the sequence of the promenade at water's edge

从上游到下游（下排由右到左）：散步道和野餐桌、从林地通往岸边、草地、全民共享的散步道、露天小咖啡馆、水岸慢跑道、钓鱼浮桥、艺术家迪迪埃·福斯蒂诺的《欺骗世界》（Trompe le monde）镜面系列作品之一

Upstream to downstream (from right to left): the promenade and picnic tables, the access to the embankments from the forest undergrowth, the meadow, the shared promenade, the café terraces, jogging at the water's edge, the pontoon of the fishermen, one of "Trick the world" mirrors of Didier Faustino

上图：从林地看向索恩河的景观
下图：河岸散步道——纤道和野餐桌

Top: View of the Saône from the undergrowth
Bottom: Promenade at water's edge, the haul road and the picnic tables

上排图：善良男孩工作室创作的充满趣味的陨石作品 / 高草丛与绿草地
下排图：河岸林地内部景观

Top: View of the playful meteorite of the artist Gentil Garçon / View of the tall grasses and the green beach
Bottom: View of the forest undergrowth

左图:罗歇塔耶城堡(远景)和莎宾娜·郎和丹尼耶尔·鲍曼创作的悬臂式楼梯——《美丽梯阶》(Beautiful steps, 近景)
中&右图:从《美丽梯阶》眺望索恩河,春天与秋天景色

Left: View of the chateau of Rochetaillée from the "Beautiful steps" of Lang and Baumann
Centre & right: The Saône from the "Beautiful steps", spring, fall

川俣正的树屋　　　　　　　　　　　　　　　　The tree house of Tadashi Kawamata

法国 洛里昂 / 2011-2015

the park and quay of the Indes Orientales
东 印 度 公 园 与 码 头

海港公共空间汇集

儒勒-菲里广场位于洛里昂市中心，介于市政府和海港之间，从西往东延展。这个长方形空间是过去船坞的所在地，1945年间受到战机轰炸之后填平了船坞，而这个暗淡无生气的绿色空间年久荒废，毫无用途，形成目前洛里昂城市肌理中一个明显的断裂空间。

然而，这块占地五公顷的土地却背负着一个使命：成为洛里昂市内具有整合作用的主要公共空间，并且作为人们举办各种活动和凯尔特音乐节的场所。在北边，印度码头广大的停车场被整治为公车道和宽阔的林荫人行道的空间，书报摊和露天咖啡座交错分布在人行道旁。位于中央的法乌耶迪克大草地经过重新改造，恢复了面向港口、港湾和大海必要的开放性和透明度。

几座"花园岛"错落其间：儿童游戏场、主题花园、休息场所……形成这座"中央公园"里的一个个小世界，而几条穿越性小径则将两侧堤岸连接了起来。在南边，罗昂码头广场可依照需要而成为停车场、游乐场和音乐节场地。在议会宫前面，原本的格洛坦停车场转变成"潮汐广场"：微微倾斜的硬质铺地形成镜面般的水池，水深仅几公分，时而涌现水雾喷洒。池水一旦流空，这个可吸收泛洪的广场便成为朝向草地开放的广大舞台。全新的东印度公园强化了洛里昂市的历史特色，同时汇集了各种具有互补用途的公共空间：公园、林荫道、广场。不论是为日常活动或集会节庆活动所用，这些充满活力的开放场所俨然与港口合为一体。

本方案继2011年设计竞赛夺标后，目前处于研究与落实的阶段。

A collection of public maritime spaces

In the heart of the city of Lorient, the present Jules-Ferry Place stretches from west to east, between the Hotel de Ville and the port. The long rectangle was formerly the wet dock, filled in after the bombardments of the Second World War. This bleak green space, aging and without use, has torn a hole in the Lorient urban fabric.

However this area of more than five hectares is slated to become the major public and unifying space of the city, for hosting demonstrations, for the Interceltic festival, as well as for daily uses. To the north, the large parking lot of the Indes quay has given way to a dedicated bus lane and to a wide tree-lined pedestrian promenade, punctuated by kiosks and café terraces. In the centre, the big Faouëdic meadow has been redone to restore the necessary transparency and opening towards the port, the harbour and the sea.

Smaller gardens, islands of green, dot the area and form microcosms in the middle of this "central park": playgrounds, theme gardens, areas for relaxation. Paths connect two old quays. To the south, the Quai de Rohan esplanade is used alternately for parking, carnivals and festivals. At the foot of the Palais des Congrès, the present Glotin parking lot is transformed into Tides Place: the slight dip in the mineral surfacing creates a reflecting pool a few centimetres thick, equipped with misting sprayers. Once empty, this floodable parvis becomes a vast stage open to the meadow. The new Indes Orientales Park strengthens the historic identity of Lorient and provides an ensemble of public spaces with complementary uses: the park, the promenade, the public square. Designed for daily use as well as for temporary events, these active and open areas unite with the port.

Following a competition won in 2011, this project is at present in the process of being studied and actualized.

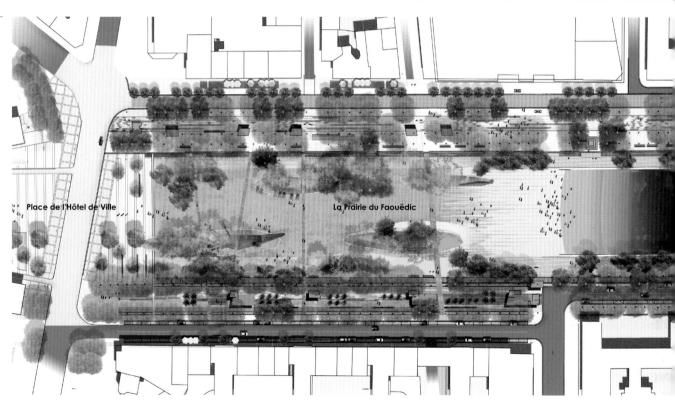

上排图：议会宫广场整治前后景观 / 从议会宫阳台看到的"潮汐广场"和法乌艾迪克草地
下图：总平面图（竞赛方案）

Top: Before and after views of the Palais des Congrès from the park / Perspective view from the balcony of the Palais des Congrès, Tides Place and the Faouëdic meadow
Bottom: "Competition" Master plan

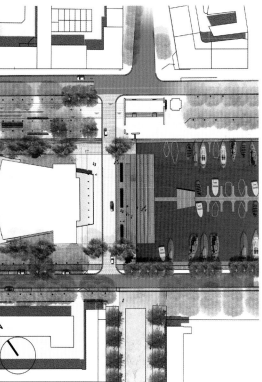

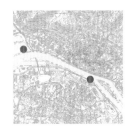

法国 鲁昂 / 2011-2015

the quays and the Waddington peninsula
沃 丁 顿 半 岛 与 码 头

跨塞纳河两岸的港口公园

多年以来，鲁昂的低码头处于废弃状态，这里的空间成为大型停车场和每年为期一个月的圣罗曼游乐场用地。而位于西侧塞纳河右岸的沃丁顿半岛，这块广大的土地也正在转变中，一些节庆活动、马戏团、音乐会和鲁昂帆船节都在这里举行。

尹西图事务所同时针对河两岸进行构思，发展出一种能提供各种不同用途的大型海港公园系统。在左岸，面向圣凯瑟琳海岸线和大教堂，沿着码头堤岸边缘伸展的散步道长达一公里，沿路种植法国梧桐树丛，几道跨河桥梁为此步道带来韵律节奏，并形成三个段落场景。位于上游处、朝向塞纳河开展的圣瑟维草地最近刚完成整治，重新改造成台阶式的堤岸让人能更亲近水面；这块河岸草地上安置了一些躺椅，丛丛柳树点缀其间。靠近下游处的克拉克东平行绿化带种植了绿意成荫的槭树林，林荫下则设置了儿童游戏场。最长的曲宏德里段落结合了宽广的草坪、杨树林和硬铺地广场；种植在箱盆中的苹果树园可依不同活动需要移动位置。一道大防波堤带人抵达码头岸边，并拥有眺望河对岸的景观视野。整个散步道设置了简约、朴实的城市小品和标识系统，体现出海港的工业特色。

在沃丁顿半岛的整治工程仅限于预先绿化阶段：排水沟渠和蓄水池的安排不仅让雨水回收措施成为景观的一部分，也重新组织了土地结构，并使这块工业废弃用地的土壤变得肥沃丰饶。围篱种植与线状植树为一个宽广的接待会场建立了架构，形成一个可以容纳临时市集或游乐场的崭新空间。福楼拜桥和测潮塔底下的镜面水池就在未来的半岛入口旁边……

A harbour park on the two banks of the Seine

Over the years, the lower quays of Rouen were abandoned to make way for a large parking structure and, for one month a year, the Saint-Romain Fair. Farther west, on the right bank, the Waddington peninsula is a vast area in transition, now used for events, circuses, concerts and the Armada of old sailing ships.

An organized brainstorming about the two banks led to the development of the framework of a large harbour park planned for different uses. On the left bank, facing Sainte-Catherine hill and the cathedral, the quayside walk stretches a kilometre; planted with pollarded plane trees and rhythmed by several bridges that form three distinct spaces. Upstream, Saint-Sever Meadows, recently redeveloped, looks over the Seine. The terraced steps of the redesigned quay allow for easy access to the water's edge. This riparian meadow, furnished with lounge-chairs, is punctuated with willow groves. Downstream, the Slipways of Claquedent, a park, offer playgrounds under the shade of maple trees. Next in the sequence is the Curanderie, blending grassy stretches, poplar groves and a stone esplanade. The orchards of apple trees are mobile, planted in wagons, to be moved in accordance with events. The big jetty provides access to the edge of the quay and offers views of the opposite side. The furniture and all the signs are plain and rough-hewn, echoing the industrial identity of the port.

In Waddington, the development has been limited to a "pre-greening": furrows left by old channels and basins direct rain-water collection, re-organize the division of land and fertilize this industrial brownfield. Plantings of hedges and aligned trees frame a vast gathering area, the new fairgrounds. At the foot of Flaubert bridge and the Marégraphe Tower, a reflecting pool adorns the entry to this peninsula in the making...

上图：河左岸低码头——被汽车占据的港口废墟以及沃丁顿半岛的整治 前后景观
下图：测潮塔和镜面水池

Top: Before and after of the lower quays of the left bank, the port wastelands occupied by cars and the Waddington peninsula
Bottom: Perspective view of the Marégraphe and the reflecting pool

上图：克拉克东平行绿化带透视图——低码头、带状花园和游戏场
下图：从"长颈鹿"观景台俯瞰曲宏德里段落的透视图，一个可以举办大型集会活动的场所

Top: Perspective view of the sequence of the "Slipways of Claquedent" from the lower quays, garden strips and games
Bottom: Perspective view of the sequence "Curanderie" from the Giraffe, a look-out tower; an open space, available for hosting large events

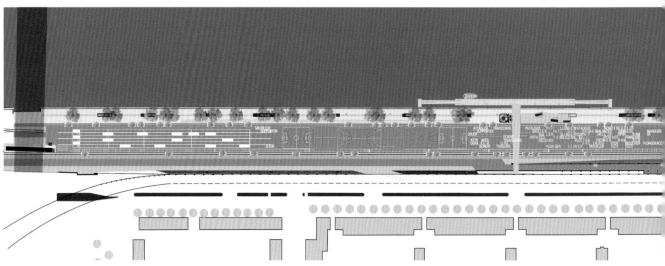

上排图：草地的夜间景观 / 邻接圣瑟维草地的一系列带状绿化园，种植着花卉与多年生禾本植物 / 沿着码头堤岸边缘伸展的散步道 / 草地上的躺椅与沿河岸伸展的小径 / 圣瑟维草地和设置其上的躺椅
下图：低码头总平面图（竞赛方案）

Top: Nocturnal view of the meadow / Strips of meadow, annuals and perennials, accompany and border the Saint-Sever meadow / View at the quay's edge / The meadow's berms and the lounge chairs / View of lounge chairs in the Saint-Sever meadow
Bottom: "Competition" Master plan of the lower quays

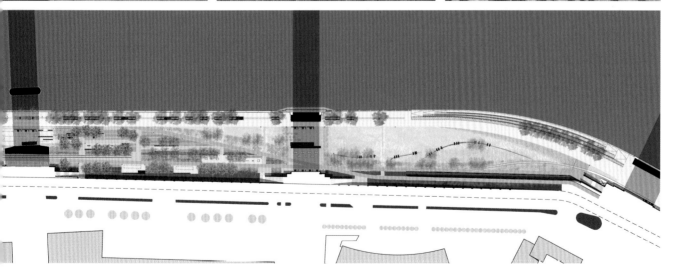

圣瑟维草地和码头堤岸拼图,项目第一阶段完成部分:阶梯平台、躺椅、水岸野餐桌、草地、旧铁轨、边界地带的绿化处理……

Mosaic of the Saint-Sever meadow at the edge of the quay, realisation of the first phase: the steps, the lounge chairs, the picnic tables at the edge of the water, the meadow, the former rails, the planted border area...

上图：圣瑟维草地，基地内的旧铁轨被保留下来，形成一条人行小径，一旁设置着带有花卉的草地
上图：从高乃依桥上眺望草地和河岸阶梯

Top: View of the Saint-Sever meadow, the existent rails are conserved and become a path bordered by the flower meadow
Bottom: View of the steps and the meadow from Corneille Bridge

projects index

方案索引

尹西图（In Situ）事务所直至2011年完成的研究与设计方案为埃曼纽尔·加勒贝尔和安妮·塔迪冯的共同作品
以下资料中的造价为不含税价格

The projects and studies realized up until 2011 are the joint effort of Emmanuel Jalbert and Annie Tardivon In Situ.
The following construction costs are calculated excluding VAT.

AGDE (France)
卡那列生态街区 Canalet Eco-neighbourhood
2013-2015, 进行中 / in progress
Atelier IN SITU (project representative)
合作者 / With : Tekhnê, P. Mourey, Gaxieu
业主 / For : CAHM
100 ha

ANNEMASSE (France)
军姆拉季广场 La place du Jumelage Jumelage Place
2010-2014
Atelier IN SITU (project representative)
合作者 / With : B. Paris, E2CA, Les Éclaireurs
业主 / For : Town Council
6.3 ha – 4.2 M€

ARGENTEUIL BEZONS (France)
塞纳河岸 The Banks of the Seine
2010-2014, 研究项目 / study
Atelier IN SITU (project representative)
合作者 / With : OGI, Biotec, LEA, Sémaphore
业主 / For : Argenteuil Bezons Agglomération Community
50 ha – 150 M€

BASEL (Switzerland)
隆多福公园 Landhof Park
2013, 竞赛项目 / competition
Atelier IN SITU (project representative)
合作者 / With : Gies architekten, T. Boutonnier artiste, Les Éclaireurs
业主 / For : Basel City Council
1 ha

BERNEX (Switzerland)
农业城市公园 Agro-Urban Park
2013, 竞赛项目 / competition
Atelier IN SITU (project representative)
合作者 / With : Tectoniques, CSD, AEU, POLEN
业主 / For : Bernex Town Council
9 ha – 6.8 M€

BOBIGNY (France)
市政府公园 Hôtel de Ville Park
2012
Atelier IN SITU (project representative)
合作者 / With : ICC, Sol Paysage, Les Éclaireurs
业主 / For : Bobigny Town Council
2.4 ha – 5.3 M€

BRON (France)
保罗皮克散步道与帕里伊街区
Paul Pic Promenade and Parilly Neighbourhood
1997
Atelier IN SITU
合作者 / With : Atelier de la Gère (project representative), Ingedia
业主 / For : Grand Lyon
5 M€

CAYENNE (France)
圭亚那纪事与文化之家
House of Memories and Cultures of Guyane
2013-2015, 进行中 / in progress
Atelier IN SITU
合作者 / With : Moreau Kusonoki (project representative), Betom, Studio A. Gardère, Tribu
业主 / For : General Council of Guyane
2 ha – 1.215 M€ (outdoor spaces)

CHAUMONT-SUR-LOIRE (France)
第八届国际花园展《只要蔬果园》
8th International Garden Festival "Nothing but Vegetable Gardens"
1999
Atelier IN SITU « Jardin Gazpatio Andaluz »
业主 / For : Garden Conservatory of Chaumont-sur-Loire

CHENÔVE (France)
克鲁奇广场与大林荫道城市公园
Colucci Place and the Urban Park of the Large Promenade
2010-2015

Atelier IN SITU (project representative)
合作者 / With : JDM, Les Éclaireurs
业主 / For : Chenove Town Council
3.5 ha – 3.8 M€

CLERMONT-FERRAND (France)
五月一日广场 First of May Place
1999-2002

Atelier IN SITU
合作者 / With : B. Paris (project representative), Sefco, JML, LEA
业主 / For : Clermont-Ferrand City Council
6 ha – 3 M€

DUBLIN (Irlande)
皇家运河河岸城市公园
Royal Canal Linear Park
2005, 国际竞赛项目 / international competition

Atelier IN SITU
业主 / For : RIAI, The Royal Institute of the Architects of Ireland
14.7 M€

FONTAINES-SUR-SAÔNE (France)
栗树街区 Marronniers Neighbourhood
1997

Atelier IN SITU (project representative)
合作者 / With : ESB, Jeol éclairagiste
业主 / For : Urban Community of Lyon
1.5 ha – 2.3 M€

FORT-DE-FRANCE (France)   pp.086-091
大拉萨瓦公园 The Large Savannah
2002-2015, 进行中 / in progress

Atelier IN SITU
合作者 / With : Olivier Dubosq architecte (project representative), LEA, GEC
业主 / For : Fort-de-France Town Council
6.7 ha – 9.45 M€

FRIBOURG (Switzerland)
红衣主教街区——蓝色工厂
Cardinal Eco-neighbourhood – Blue Factory
2013, 竞赛项目 / competition

Atelier IN SITU
合作者 / With : FHY (project representative), CSD, CITEC
业主 / For : Fribourg Town Council
6 ha

GENEVA (Switzerland)
巴提森林游戏场 Bâtie Woods Play Area
2013, 竞赛第二名 / runner-up in the competition

Atelier IN SITU (project representative)
合作者 / With : Ganz & Muller, Caracol
业主 / For : Geneva City Council
1.5 ha

GIF-SUR-YVETTE (France)
穆隆公园，巴黎-萨克雷大学校园
Moulon Park, Paris-Saclay Campus
2015, 进行中 / in progress

Atelier IN SITU (project representative)
合作者 / With : Encore Heureux, F. Magos, Infraservices, Sol paysage, REP
业主 / For : Paris-Saclay Public Establishmen, 4 ha – 9 M€

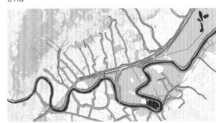

GRENOBLE (France)
格勒诺布尔城东极城区发展研究
Study for the East Polarity of Grenoble
2013-2015, 进行中 / in progress

合作者 / With : XDGA urbanistes (project representative), REP
业主 / For : Grenoble Alpes Métropole
750 ha

GRUYÈRES (Switzerland)
停车场 Parking Areas
2014-2015

Atelier IN SITU
合作者 / With : Joud & Vergely architectes (project representative)
业主 / For : Gruyères Town Council
10 ha – 2.5 M€

HAMBURG (Germany)
港口新城 Hafencity – the Port City
2000, 国际竞赛项目 / international competition

Atelier IN SITU
合作者 / With : F.H. Jourda (project representative)
业主 / For : Hambourg City Council
155 ha

HELSINKI (Finland)
西港公园 Länsisatamanpuisto Park
2005, 竞赛第三名 / 3rd prize in the international competition

Atelier IN SITU (project representative)
合作者 / With : LEA
业主 / For : Helsinki city Planning Department
1.05 ha

LA ROCHELLE (France)
弥和耶街区 Mireuil Neighbourhood
2007-2015
Atelier IN SITU (project representative)
合作者 / With : Rue Royale, Agathe Argod, ICC
业主 / For : La Rochelle City Council
4 ha – 5 M€

LAUSANNE, RENENS, PRILLY (Switzerland) pp.024-027
马利街区平行绿化带 The Corridors of Malley
2011-2015, 进行中 / in progress
Atelier IN SITU (project representative)
合作者 / With : FHY, CSD
业主 / For : SDOL, Renens and Prilly Town Council
50 ha – 14 M€

LAUSANNE (Switzerland)
勒福隆街区 Le Flon Neighbourhood
2015, 竞赛项目 / competition
Atelier IN SITU (project representative)
合作者 / With : CSD, Les Éclaireurs, Ruedi Baur
业主 / For : MOBIMO, Lausanne
2 ha – 8 MCHF

LORIENT (France) pp.112-115
东印度公园与码头
The Park and Quay of the Indes Orientales
2011, 进行中 / in progress
Atelier IN SITU (project representative)
合作者 / With : CMJN, Artelia, Les Éclaireurs
业主 / For : Lorient Town Council
5.6 ha – 5 M€

LYON 1st (France) pp.066-071
圣让堡 Fort Saint-Jean
1999-2005
Atelier IN SITU
合作者 / With : Vurpas et associés (project representative), E2CA
业主 / For : Ministry of Economy and Finance
1 ha – 1 M€

LYON 3rd (France)
夏尔梅特花园 Charmettes Garden
1996
Atelier IN SITU (project representative)
合作者 / With : BREA
业主 / For : Lyon City Council
1 500 m² – 1.5 MF

LYON 5th (France)
瓦隆装广场 Valensio Place
1998-2002
Atelier IN SITU (project representative)
合作者 / With : Sefco, Agibat, Antea
业主 / For : Lyon City Council, DETU
1 850 m² – 0.76 M€

LYON 3rd, 6th, 7th (France) pp.092-103
罗纳河岸 The Banks of the Rhône
2003-2008
Atelier IN SITU (project representative)
合作者 / With : F.H. Jourda, Coup d'Éclat, GEC, ARTELIA, Biotec, Agibat, Sol paysage, P. Favier
业主 / For : Grand Lyon
10 ha – 30 M€

LYON 8e (France)
杰尔兰路75号生态街区 75 Rue de Gerland Eco-neighbourhood
2011-2015, 进行中 / in progress
Atelier IN SITU (project representative)
合作者 / With : Garcia Diaz architectes, Betrec
业主 / For : D2P, Gecina
5 000 m² – 1.2 M€

LYON 8th (France) pp.016-023
1945年5月8日广场 8 May 1945 Place
1998-2002
Atelier IN SITU (project representative)
合作者 / With : F.H. Jourda, Sefco, LEA, JML, Pierre Pardon sculpteur
业主 / For : Grand Lyon
3 ha – 5 M€

MÂCON (France) pp.046-051
马贝街区散步道公园 The Park Walk of Marbé
2007-2014
Atelier IN SITU (project representative)
合作者 / With : E2CA, ARBOR&SENS
业主 / For : Mâcon Town Council
51 ha – 22 M€

MANNHEIM (Germany)
BUGA 2023城市公园 BUGA Park 2023
2014
Atelier IN SITU (project representative)
合作者 / With : Zeppelin
业主 / For : Stadt Mannheim, Fachbereich für Stadtplanung
50 ha

MARSEILLE (France)
圣让堡花园 Fort Saint-Jean Gardens
2010, 竞赛项目 / competition
Atelier IN SITU (project representative)
合作者 / With : OGI, Les Éclaireurs, Bertrand Retif botaniste
业主 / For : Ministry of Culture
1.5 ha – 5.7 M€

MEUDON, SÈVRES, ISSY-LES-MOULINEAUX (France)
塞纳河岸 The Banks of the Seine
2006, 项目定位研究 / preliminary project definition competition
Atelier IN SITU (project representative)
合作者 / With : Sogreah, Biotec, Marchetto'Ecco, LEA, Isabelle Hurpy
业主 / For : General Council of Hauts-de-Seine
23 ha – 28 M€

MONTPELLIER (France)
拉黑斯唐克街区 La Restanque Neighbourhood
2007, 竞赛项目 / competition
Atelier IN SITU
合作者 / With : Atelier Philippe Madec (project representative)
业主 / For : Montpellier City Council
140 ha – 85 M€

MULHOUSE (France)
DMC生态街区 DMC Eco-neighbourhood
2009-2010
Atelier IN SITU
合作者 / With : Atelier B. Paris (project representative)
业主 / For : Mulhouse City Council
70 ha – 47 M€

NANTES (France)
南特岛北河岸 The North Banks of the Isle of Nantes
2013, 竞赛项目 / competition
Atelier IN SITU (project representative)
合作者 / With : FHY, AIA, Biotec, Collectif FTC, Les Éclaireurs
业主 / For : SAMOA
7 ha – 7 M€

NICE (France)
神殿喷泉广场 Temple Fountain Place
2003-2008
Atelier IN SITU
合作者 / With : OTH, LEA
业主 / For : Nice City Council
1.5 ha – 3.1 M€

NIORT (France)
城堡主塔广场 Donjon Place
2006-2014
Atelier IN SITU (project representative)
合作者 / With : Atelier Lion, François Magos, E2CA
业主 / For : Niort Town Council
8 675 m² – 3 M€

NOISY-LE-SEC (France)
乌尔克平原可持续发展街区
Sustainable Neighbourhood of the Ourcq Prairie
2014-2015, 进行中 / in progress
Atelier IN SITU (project representative)
合作者 / With : ECCTA, VERDI Seine
业主 / For : Sequano Aménagement
3.8 ha – 15 M€

ORLÉANS (France)
卢瓦尔河码头 Loire Quays
2011, 竞赛项目 / competition
Atelier IN SITU (project representative)
合作者 / With : FHY, Sogreah, Les Éclaireurs
业主 / For : Orléans City Council
4.5 M€

PARIS 13th (France)
萨勒贝提耶通道 Salpetrière Open Space
1999-2003
Atelier IN SITU (project representative)
合作者 / With : Segic
业主 / For : SEMAPA
800 m² – 1.7 M€

PARIS 14th (France)
引水道花园 Aqueduct Garden
1999-2001
Atelier IN SITU (project representative)
合作者 / With : SEFCO
业主 / For : Paris City Council DPJEV
2 100 m² – 1.2 M€

PARIS 18th (France)
罗莎·卢森堡花园 Rosa Luxemburg Garden
2007-2014
Atelier IN SITU
合作者 / With : F.H. Jourda (project representative), Infra Services, LEA, Sol Paysage
业主 / For : Paris City Council
9 500 m² – 30 M€ (total amount), 3 M€ (gardens)

pp.052-059

REIMS (France)
兰斯2020发展展望研究，可持续发展的大都会
Reims 2020 prospective study, sustainable metropolis
2009-2010
Atelier IN SITU
合作者 / With : Devillers & associés (project representative)
业主 / For : Reims Métropole

RENNES (France)
圣安妮广场 Sainte-Anne Place
2014, 竞赛项目 / competition
Atelier IN SITU (project representative)
合作者 / With : Sephia, ECL studio, Tekhnê
业主 / For : Rennes City Council
1.2 ha – 3.5 M€

ROCHETAILLÉE-SUR-SAÔNE (France) pp.104-111
小咖啡馆散步道 The Promenade of the Cafés
2011-2014
Atelier IN SITU (project representative)
合作者 / With : Sinbio, ICC, LEA, OGI
业主 / For : Grand Lyon
6 ha – 5.2 M€

ROMANS-SUR-ISÈRE (France) pp.028-031
韦科尔大型散步道
The Great Promenade of Vercors
2013-2014
Atelier IN SITU (project representative)
合作者 / With : Tekhnê, Infraservices, Atelier Lumière, ICC
业主 / For : Romans-sur-Isère Town Council
5 ha – 8 M€

ROUEN (France) pp.116-123
沃丁顿半岛与码头
The quays and the Waddington Peninsula
2011-2015, 进行中 / in progress
Atelier IN SITU (project representative)
合作者 / With : FHY, Artelia, Les Éclaireurs, Sol Paysage, C&E
业主 / For : Rouen City Council
15 ha – 17.3 M€

SAINT-CHAMOND (France) pp.072-075
炼钢厂演化公园
The Changeable Park of the Aciéries
2010-2012
Atelier IN SITU (project representative)
合作者 / With : Réalité Les Éclaireurs
业主 / For : Saint-Étienne Métropole
1.5 ha – 1.8 M€

SAINT-CHRISTOL (France) pp.076-079
维阿维诺葡萄品种栽培园
The Ampelographic Gardens of Viavino 2008-2013
Atelier IN SITU
合作者 / With : Atelier Madec (project representative), MC Pro, Tribu, ICC
业主 / For : Lunel Country Municipal Community Council
2.35 ha – 1 M€

SAINT-ÉTIENNE (France)
城市大街 Great Street
1997-2000
Atelier IN SITU (project representative)
合作者 / With : E2CA, BET VRD
业主 / For : SIOTAS
1.3 km – 7 M€

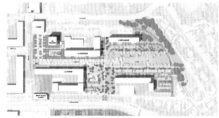

SAINT-ÉTIENNE (France)
蒙黑诺街区 Montreynaud Neighbourhood
2011-2015, 进行中 / in progress
Atelier IN SITU (project representative)
合作者 / With : Pierre Scodellarri architecte, E2CA, Cetis
业主 / For : Saint-Étienne City Council
20 ha – 5.6 M€

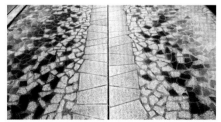

SAINT-FLOUR (France)
广场与街道 City Squares and Streets
1998-2006
Atelier IN SITU (project representative)
合作者 / With : ICC, LEA
业主 / For : Saint-Flour Town Council
1.6 ha – 3.1 M€

SAINT-MARTIN-D'HÈRES (France) pp.062-065
露西·欧布拉克广场 Lucle Aubrac Place
2005-2007
Atelier IN SITU (project representative)
合作者 / With : E2CA
业主 / For : GPV Grenoble, Saint-Martin-d'Hères Town Council
0.8 ha – 1.5 M€

SAINT-PRIEST (France)
贝莱尔散步道 Bel Air Promenade
1999-2002
Atelier IN SITU (project representative)
合作者 / With : BREA
业主 / For : Grand Lyon, service DSQ
4.2 ha – 2.8 M€

SAINT-PRIEST (France)   pp.080-083
纳尔逊·曼德拉公园 Nelson Mandela Park
2013-2014

Atelier IN SITU (project representative)
合作者 / With : Sotrec, AEU, Les Éclaireurs, Soberco, E. Roule
业主 / For : Saint-Priest Town Council
12 ha – 2.8 M€

SION (Switzerland)
穿越罗纳河 Rhone Crossing
2013, 竞赛项目 / competition

Atelier IN SITU
合作者 / With : P. Gagliardi (project representative), Nomad architecte, CSD
业主 / For : Sion Town Council
12.8 M€

STRASBOURG (France)
城堡广场 Château Place
2011, 竞赛项目 / competition

Atelier IN SITU (project representative)
合作者 / With : Sogreah, F. Magos, P. Favier artiste
业主 / For : Strasbourg City Council
6.5 ha – 1.4 M€

TOULOUSE (France)
撒拉德生态街区 Salade Neighbourhood
2013-2015, 进行中 / in progress

Atelier IN SITU
合作者 / With : Tekhnê (project representative), 2AU, C2I, Soberco
业主 / For : Habitat Toulouse
3 ha – 2 M€

TRÉVOUX (France)
火车站生态街区 Train Station Eco-neighbourhood
2014-2015, 竞争性对话 / competitive dialogue

Atelier IN SITU
合作者 / With : Tekhnê (project representative), ICC, Soberco
业主 / For : Trévoux Town Council
9 ha – 5.1 M€

TURIN (Italy)
皮雷利街区 Pirelli Neighbourhood
2012, 竞赛项目 / competition

Atelier IN SITU
合作者 / With : Picco architetto (project representative), Crotti & Forsans, Inpro
业主 / For : Pirelli
15 ha

VAL-DE-REUIL (France)
挪亚生态村落 Noës Ecovillage
2009

Atelier IN SITU
合作者 / With : Quille (project representative), Atelier Philippe Madec, Tribu
4 ha

VAULX-EN-VELIN (France)
园区公园 Campus Park
2015, 进行中 / in progress

Atelier IN SITU
合作者 / With : Urbino architecte (project representative), Artelia, Les Éclaireurs
业主 / For : ENTPE, ENSAL
5 ha – 4 M€

VAULX-EN-VELIN (France)   pp.034-037
提伯德街区 La Thibaude Neighbourhood
1996-1997

Atelier IN SITU (project representative)
合作者 / With : Bruno Dumétier, BREA
业主 / For : Grand Lyon
4 ha – 3.5 M€

VÉNISSIEUX (France)
普左兹街区 Puisoz Neighbourhood
2009-2015, 进行中 / in progress

Atelier IN SITU
合作者 / With : Atelier T. Roche (project representative), Sotrec
业主 / For : Lionheart, D2P
20 ha – 15 M€

VILLEURBANNE (France)   pp.012-015
佩里松小区花园 The Gardens of Pélisson City
1997

Atelier IN SITU (project representative)
业主 / For : Villeurbanne City Council
0.84 M€

VILLEURBANNE (France)   pp.038-045
拉扎尔-古戎广场 Lazare-Goujon Place
2004-2008

Atelier IN SITU (project representative)
合作者 / With : P. Favier artiste, E2CA, Coup d'Éclat, JML
业主 / For : Grand Lyon
1 ha – 3.1 M€

experience 事务所经历

尹西图（In Situ）事务所于1991年成立于里昂，是埃曼纽尔·加勒贝尔和安妮·塔迪冯联合建立的工作室，两人都是毕业于凡尔赛国立高等景观设计学院的景观建筑师，并且在合伙创办事务所之前，个别跟随米歇尔·寇拉儒、亚历山大·谢梅道夫、阿兰·马格里等知名的景观界前辈历练了几年。

当时里昂对创造城市公共空间的积极政策使尹西图事务所得以在里昂城市共同体的地域范围内实现最早几个具有代表意义的项目，特别是一些早期建造的大型社会住宅区的更新项目，以及一些较小城乡的整治项目。事务所在维勒班市为佩里松社会住宅小区所进行的更新整治方案获得了1997年的法国景观锦标奖，此后展开了几项较大规模的重要城市项目：里昂市的1945年5月8日广场、维勒班市的拉扎尔-古戎广场、里昂市的罗纳河岸整治、巴黎市的罗莎·卢森堡花园……事务所逐渐茁壮、扩充，并聘请了多位项目设计经理。埃曼纽尔·加勒贝尔和安妮·塔迪冯除了项目设计之外，也同时从事教学工作，并担任政府机构的景观顾问。

2008年的夏天，一场意外火灾将事务所刚迁入的办公室全然烧毁，而后两位合伙领导人之间也相继出现重要的意见分歧，因此两人在2011年签下拆伙协议，决定由埃曼纽尔·加勒贝尔继续领导事务所，并和里昂办公室的专业同仁共同完成进行中的所有项目。安妮·塔迪冯则于2012年在巴黎创立了Inuits*事务所。

转变后至今，事务所维持在十人左右的规模，包括景观设计师、工程师和城市规划师，主要进行位于法国境内与海外领土，以及欧洲的项目。事务所不仅进行景观设计与实施的工作，也参与城市研究和拟定发展计划的任务，以景观设计师和城市规划师的身份承揽跨领域的项目，扮演项目总承包的角色，积极进行生态街区规划以及可持续发展的城市更新方案。

事务所也在项目的进行中展开各类民众参与的程序，包括与使用者、周围居民以及管理单位的积极协商。尹西图事务所在创造共享花园的领域上发展出扎实的专业知识与技能，同时也对于城市农业发展与创新、永久性农业以及替代式雨水管理等范畴相当投入。此外，事务所在项目构思阶段便致力于将整治空间日后的维护与保养问题纳入考量。尹西图事务所经常与不同领域的合作单位、具有互补专长的团队共同构思方案，以便打造具有创新精神、多样而开放的城市方案。

*注释：根据双方签订的协议，安妮·塔迪冯完全可以引用尹西图（In Situ）事务所直至2011年完成的研究与设计方案作为其工作经历的参考，但必须严格地以个人身份引用。无论是其新创立的Inuits公司，或者其合伙人文森·贝纳尔，皆无权引用这些并不属于他们的工作成果。

左页：吉约蒂耶尔水池，罗纳河岸，里昂

The agency In Situ was founded in Lyon in 1991, by Emmanuel Jalbert and Annie Tardivon, both graduates of the École Nationale Supérieure du Paysage de Versailles (the National Graduate School of Landscaping), after a preliminary cycle in Architecture. They each collaborated for several years with Michel Corajoud, Alexandre Chemetoff and Alain Marguerit.

The policy of the creation of public spaces in which they engaged during this time allowed the agency to realise its first significant projects in the greater urban area, through urban renewal projects both in large housing estates. In 1997 the Landscaping Trophy was given to In Situ for the redevelopment of the Pélisson housing estate in Villeurbanne. Following came developments of greater scope: 8 May 1945 Place in Lyon, Lazare-Goujon Place in Villeurbanne, the embankments of the Rhône in Lyon, Rosa Luxemburg Garden in Paris... The agency is growing, expanding, and calls upon the services of several project leaders. At the same time, Annie Tardivon and Emmanuel Jalbert conducted teaching and landscaping advisory activities for the government.

The path of the two founding partners diverges in 2008. Since 2011, Emmanuel Jalbert has pursued the work of In Situ, with all of the project leaders and employees in Lyon, in the framework of the In Situ studio / Landscapes and Urban Planning. Annie Tardivon has led the Inuits* in Paris since 2012.

Today, the In Situ studio has brought together a dozen collaborators, landscapers, engineers and urban planners. The studio works in France, Europe and over-seas, through project management assignments but also, preliminarily, through urban studies both as landscapers and urban planners. Agents of multidisciplinary teams, they work on projects of eco-districts and of sustainable urban renewal.

The studio has also invested in various participatory approaches, integrating an active dialogue with site users, river inhabitants and managers. The studio has also pursued a committed approach to the development and innovation in the fields of, urban agriculture, permaculture and alternative management of rainwater. It is committed to incorporate at the conception of the project, the upkeep and maintenance of developed spaces. In Situ collaborates regularly with numerous partners, in the framework of innovative, expressly diverse and open, urban projects.

* Note : In conformity with the signed agreement, Annie Tardivon can personally use references to the studies and projects realised with In Situ up until the end of 2011. On the other hand, neither the Inuits company nor the other partners of this society hold the rights to use these references that in no way belong to them.

Opposite page: The basin of the Guillotière, les banks of the Rhône, Lyon

team

2015年团队：

Emmanuel Jalbert（埃曼纽尔·加勒贝尔）, Anne Romettino, Fabrice Lazert, Yann Chabod, Anna Thomé, Simon Kuntze-Fechner, Emilie Collavet, Julien Baby, Monique Melmoux.

前合伙人（直到2011年）：Annie Tardivon（安妮·塔迪冯）

从前工作伙伴：

Anna-Andréa Obé-Gervais, Suzanne Chatelard, Marie-Gabrielle Beuvier, Michael Rosso, David Schulz, Morgane Le Bissonais, Julien Petiot, Marion Vittupier, Joseph Marche, Eve Marre, Rudy Toulotte, Pierre Derycke, Alexis Delekta, Frédéric Agnésa, Mounira Athmani, Vincent Bénard, Jeanne Bouët, Grégory Cluzel, Laurence De Lorenzi, Laurence Deschaux, Philippe Diaz, Nathalie Drevet, Olivier Fayolle, Claire Flandin, Yannis Le Quintrec, David Maire, Mahaut Michez, Cécile Paris, Marc Pelosse, Sophie Peyrard, Marc Raynaud, Frédéric Reynaud, Pierre Rotival, Pierre Scodellari, Coralie Scribe, Mélanie Tant, Christophe Veyrat-Parisien, Pascale Acciari.

Curent team, in 2015:

Emmanuel Jalbert, Anne Romettino, Fabrice Lazert, Yann Chabod, Anna Thomé, Simon Kuntze-Fechner, Emilie Collavet, Julien Baby, Monique Melmoux.

Former partner: Annie Tardivon, up until 2011

Former collaborators:

Anna-Andréa Obé-Gervais, Suzanne Chatelard, Marie-Gabrielle Beuvier, Michael Rosso, David Schulz, Morgane Le Bissonais, Julien Petiot, Marion Vittupier, Joseph Marche, Eve Marre, Rudy Toulotte, Pierre Derycke, Alexis Delekta, Frédéric Agnésa, Mounira Athmani, Vincent Bénard, Jeanne Bouët, Grégory Cluzel, Laurence De Lorenzi, Laurence Deschaux, Philippe Diaz, Nathalie Drevet, Olivier Fayolle, Claire Flandin, Yannis Le Quintrec, David Maire, Mahaut Michez, Cécile Paris, Marc Pelosse, Sophie Peyrard, Marc Raynaud, Frédéric Reynaud, Pierre Rotival, Pierre Scodellari, Coralie Scribe, Mélanie Tant, Christophe Veyrat-Parisien, Pascale Acciari.

interventions-distinctions

活动与得奖记录

得奖

- 2014年城市艺术奖所颁发的市长奖（罗伯特·乌泽勒研讨会），得奖项目：大拉萨瓦公园，位于法兰西斯堡
- 2014年景观优胜奖，马孔市与马孔住宅局的专业合作对象，得奖项目：马尔贝街区散步道公园
- 2013年罗纳河建筑、城市与环境大奖的《城市与景观整治》佳作奖，得奖项目：拉扎尔-古戎广场，位于维勒班市
- 2012年景观优胜奖，大里昂区域《空间与时间》特别奖，得奖项目：罗纳河岸与河畔公园
- 2008年Le Moniteur出版集团城市整治锦标奖的特别佳作奖，得奖项目：罗纳河岸
- 2007年城市入口奖，得奖项目：大拉萨瓦公园，位于法兰西斯堡
- 2003年Le Moniteur出版集团的城市整治奖，得奖项目：1945年5月8日广场
- 2002年法国电力集团设计竞赛首奖，得奖项目：大拉萨瓦公园
- 1997年法国景观锦标奖，法国政府地域与环境整治部颁发，得奖项目：佩里松小区花园，位于维勒班

展览

- 《索恩河畔》，法国里昂，2011
- 《水工程》，加州艺术学院，美国旧金山，2011
- 《丰饶城市，景观的制造》，建筑与文化遗产中心，法国巴黎，2011
- 《流动的城市》，设计双年展，法国圣艾蒂安，2010
- 《生态居住》，阿森纳展览馆，法国巴黎，2009
- 《罗纳河岸，生活的岸地、抵达的岸地》，法国里昂，2007
- 《里昂罗纳河畔低码头整治项目》，AFA展览，法国观点，中国北京、上海、广东，2005
- 《移动建筑》，巡回展览，墨西哥、巴黎、阿姆斯特丹、横滨、上海等等，2004
- 《地域》，USA景观建筑师，美国哈佛、加拿大蒙特利尔，2001

讲座、设计工作室与教学

- 《动态城市与景观》，罗纳河地区建筑、城市与环境协会培训课程，2015
- 《城市公义与公共空间》，里昂城市规划研究所论坛，2014
- 《索恩河岸，小咖啡馆散步道》，巴塞罗纳国际景观双年展，2014
- 《城市-公园》，国际公园节中的法国-俄罗斯论坛，莫斯科，2013
- 《蓝色网络》，与瑞士景观建筑师联合协会以及IBA BASEL 2020合作举办的景观研讨大会，2013
- 《马利街区平行绿化带》，EPFL建筑学院讲座，瑞士洛桑，2011
- 教学与设计工作室，凡尔赛国立高等景观设计学院、昂热园艺研究所、日内瓦吕利耶工程学院，以及里昂、格勒诺布尔和蒙特利尔等建筑学院。
- 各种设计竞赛与奖项的评审委员

出版

- 《里昂罗纳河岸，城市画像》，Stéphane Baches出版社，2009
- 其他出版，请查询本事务所网站

Awards

- Urban Art Prize 2014, Robert Auzelles Seminary, the Savannah, Fort-de-France
- Landscape Victories 2014, professional partner of the City of Mâcon and Mâcon Habitat, winner for the park walk of the Marbé neighbourhood
- Grand Prize of Architecture, Urbanism and the Environment of the Rhône 2013, mention for "urban development and landscape" Lazare-Goujon Place, Villeurbanne
- Landscape Victories 2012, special prize "Space and Time" for Great Lyon, the banks of the Rhône and the Park of the Banks
- Le Moniteur Trophy of Urban Development 2008, special mention for the Banks of the Rhône, Lyon
- Prize for City Entrances 2007, the Savannah, Fort-de-France
- Le Moniteur Trophy of Urban Development 2003, 8 May 1945 Place, Lyon
- Competition EDF 2002, winning project the Savannah
- Landscape Trophy, Minister of Development of the Territory and the Environment 1997, Pélisson City, Villeurbanne

Exhibitions

- "Rives de Saône" ("Banks of the Saône"), Lyon, 2011
- "Water works", California College of the Arts, San Francisco, 2011
- "Ville fertile, la fabrique du paysage" ("Fertile city, the making of the landscape"), City of Architecture and Heritage, Paris, 2011
- "Ville mobile" ("Mobile City"), Design Biennial, Saint-Étienne, 2010
- "Habiter écologique" ("Ecological Living"), Pavilion of the Arsenal, Paris, 2009
- "Les berges du Rhône, la rive à (re)vivre, la rive à arrive" ("The Banks of the Rhône, banks to (re)live, banks to arrive"), Lyon, 2007
- "Le projet des Bas-ports du Rhône à Lyon" ("The Project of the Low Ports of the Rhône in Lyon"), AFA exhibition, French Vision, Beijing, Shanghai, Guangdong, 2005
- "Bouge l'architecture" ("Moving Architecture") Itinerant exhibition, Mexico, Paris, Amsterdam, Yokohama, Shanghaï, etc., 2004
- "Territoires" ("Territories"), landscape architect USA, Harvard, Montreal, 2001

Conferences, Workshops & Teaching

- "Ville et paysage en mouvement" ("City and landscape in movement"), training at the CAUE of the Rhône, 2015
- "Équité urbaine et espaces publics" ("Urban Equity and Public Spaces"), forum at the Institute of Urbanism of Lyon, 2014
- "Rive de Saône, la promenade des Guinguettes" ("Banks of the Saône, the Dance-Hall Promenade"), VIIIth International Biennial of Landscaping of Barcelona, 2014
- "Ville-Parc" ("City-Park"), Franco-Russian Forum of the International Festival of Public Parks, Moscow, 2013
- "La Maille bleue" ("Blue Mesh"), Landscaping Congress organised in collaboration by the Switzerland Federation of landscape architects and IBA BASEL 2020, 2013
- "Les coulisses de Malley" ("The Corridors of Malley"), lecture at Lausanne, School of Architecture EPFL, 2011
- Teaching, workshops at the ENSP of Versailles, INH of Angers, Lullier EIL of Geneva, the Schools of Architecture of Lyon, Grenoble and Montreal
- Various competition and prize juries

Publications

- "Les berges du Rhône à Lyon, portrait d'une ville" ("The Banks of the Rhône in Lyon, portrait of a city"), éditions Stéphane Baches, 2009
- Other publications: cf. internet site

版权说明

credits

文字、照片与各种图面资料：
© IN SITU paysages & urbanisme

以下资料除外

本书章节开篇处的引用文字截取于《景观，始于天空与大地相触的地方》，作者Michel Corajoud，出版社Actes Sud，以及《北纬度，崭新城市景观》（Latitude Nord景观事务所作品专辑），作者Annette Vigny，出版社Actes Sud Nature

照片
Sébastien Delmer – pp.122-123
Christine Delpal – p.18 左下&右下, p.20 左中&右下, p.22 上, p.23 右
Olivier Dubosq – p.88 左中
Gérard Dufresne – p.17, p.19 左下, p.20 中上 p.99 上, p.100 右上
Patrick Georget – p.47
Go Production, Communauté de Communes du Pays de Lunel – p.78 左上, p79 上
Gilles Michallet – p.41 上
Minéral Service – p.120 夜景照片, p.121 左上

三维效果图
Marc Boudier – p.73, pp.118-119
CMJN architectes – p.113, p.115 上
Cyrille Jacques – p.29, p.30 上, p.31 上
Philippe Martyniak – p.81, p.83, p.89 上

方案模型
Atelier Fau – p.25

Textes, photographies, 3D models and sketches:
© IN SITU paysages & urbanisme

expect

Quotations taken from the books "Le paysage, c'est l'endroit ou le ciel et la terre se touchent" (Landscape is the place where sky and earth touch), Michel Corajoud, Actes Sud and "Latitude nord, nouveaux paysages urbains" (Northern Latitudes, New Urban Landscapes), Annette Vigny, Actes Sud Nature.

Photographies
Sébastien Delmer – pp.122-123
Christine Delpal – p.18 bottom left & bottom right, p.20 left centre & bottom right, p.22 top, p.23 right
Olivier Dubosq – p.88 left centre
Gérard Dufresne – p.17, p.19 bottom left, p.20 top centre, p.99 top, p.100 top right
Patrick Georget – p.47
Go Production, Communauté de Communes du Pays de Lunel – p.78 top left, p79 top
Gilles Michallet – p.41 top
Minéral Service – p.120 night photos, p.121 top left

3D models
Marc Boudier – p.73, pp.118-119
CMJN architectes – p.113, p.115 top
Cyrille Jacques – p.29, p.30 top, p.31 top
Philippe Martyniak – p.81, p.83, p.89 top

Project models
Atelier Fau – p.25

致谢

acknowledgements

唯有具备高度意愿并懂得信任专业的民选代表和业主，才能成功促使方案的实现；唯有热情洋溢并积极负责的设计团队和工作搭档，才能完成高品质的设计方案；唯有预先与使用者对话、倾听他们的心声，才能建造真正具有实用性的空间；唯有拥有团结一致而能力互补的合作伙伴，才能落实生态责任；唯有具有实力与创造力的植物苗圃和建设工程企业，才能造就典范项目；唯有积极进行维护管理的服务单位，以及尊重环境的使用者，才能达成公共空间的可持续发展。本人在此诚挚感谢这些实践项目的所有参与者和行动者。

感谢以下企业对本书出版的支持：Greenstyle, Grepi, Laquet, Razel-Bec, Sols Confluence, Soupe, Tarvel, Vallois, pour leur participation.

此外，我也特别感谢艾米莉·寇拉维具耐心而有建设性的贡献，以及尼古拉斯·布里左和简嘉玲（法国亦西文化公司 ICI Consultants）专业而中肯的建议。

埃曼纽尔·加勒贝尔

There are no finished projects without elected officials and volunteer clients who know how to trust, no projects of quality without a team of project contractors and enthusiastic, engaged colleagues, no developments easy to appropriate without a dialogue and a preliminary meeting with the "use manager", no eco-responsibility without united partners with complementary skills, no exemplary realisations without competent and innovative nurseries and garden suppliers, no sustainable public space without services motivated to maintain them and users who respect them and make them come alive; we wish to warmly thank all who worked on these projects, the ones that have been realized as well as the others.

Thanks to companies : Greenstyle, Grepi, Laquet, Razel-Bec, Sols Confluence, Soupe, Tarvel, Vallois, for their participation.

To Émilie Collavet for her constructive and patient contribution. And to Nicolas Brizault and Chia-Ling Chien for their considered advice in terms of the creation of this publication.

Emmanuel Jalbert

右页：吉约蒂耶尔水池色缤纷的池底，罗纳河岸，里昂

Opposite page: At the bottom of the basin of the Guillotière, les banks of the Rhône, Lyon